Modern Researches in Radioanalytical and Imaging Techniques

Modern Researches in Radioanalytical and Imaging Techniques

Edited by **Marcia Dion**

LANRYE
INTERNATIONAL

New Jersey

Published by Clanrye International,
55 Van Reypen Street,
Jersey City, NJ 07306, USA
www.clanryeinternational.com

Modern Researches in Radioanalytical and Imaging Techniques
Edited by Marcia Dion

International Standard Book Number: 978-1-63240-367-4 (Hardback)

Printed in the United States of America.

Contents

Preface

Every book is a source of knowledge and this one is no exception. The idea that led to the conceptualization of this book was the fact that the world is advancing rapidly; which makes it crucial to document the progress in every field. I am aware that a lot of data is already available, yet, there is a lot more to learn. Hence, I accepted the responsibility of editing this book and contributing my knowledge to the community.

In this book, modern researches in radioanalytical and imaging techniques are primarily elucidated. It focuses on providing the readers with rich content in the field of radioanalytic methods, particularly stressing on the growth and discoveries in this rapidly developing area. The book talks about the fundamentals and use of these methods in a vast range of science, technology and medicines, where radioanalytic modes are expected to bring considerable changes. The book also provides an outlook on the background of this area, its recent developments and the future outcomes. This text basically highlights novel applications and related tools and schemes, such as instrumentation systems and computing hardware/software. This book mainly focuses on learners, discoverers, analysts and workers who show a great interest in medical and ground penetrating radar (GPR) imaging, and radio-analytical schemes.

While editing this book, I had multiple visions for it. Then I finally narrowed down to make every chapter a sole standing text explaining a particular topic, so that they can be used independently. However, the umbrella subject sinews them into a common theme. This makes the book a unique platform of knowledge.

I would like to give the major credit of this book to the experts from every corner of the world, who took the time to share their expertise with us. Also, I owe the completion of this book to the never-ending support of my family, who supported me throughout the project.

Editor

Medical and GPR Imaging Techniques: Principles, Applications and Safe Utilizations

Principles and Applications of Nuclear Medical Imaging: A Survey on Recent Developments

Faycal Kharfi

Additional information is available at the end of the chapter

1. Introduction

The main difference between nuclear imaging and other radiologic tests is that nuclear imaging assesses how organs function, whereas other imaging methods assess anatomy, or how the organs look. The advantage of assessing the function of an organ is that it helps physicians make a diagnosis and plan present or future treatments for the part of the body being evaluated. Fast improvements in engineering and computing technologies have made it possible to acquire high-resolution multidimensional nuclear images of complex organs to analyze structural and functional information of human physiology for computer-assisted diagnosis, treatment evaluation, and intervention. Technological inventions and developments have created new possibilities and breakthroughs in nuclear medical diagnostics. The classic example is the discovery of Anger, fifty six years ago. The application and commercial success of new nuclear imaging methods depends mainly on three primary factors: sensitivity, specificity and cost effectiveness. The first two determine the added clinical value, in comparison with existing medical imaging methods. Nowadays, much greater importance is attached to cost effectiveness than in the past. This also holds true for diagnostic equipment where, for example, one of the consequences is that price erosion will occur where the functionality of an instrument is not open to further development. Cost effectiveness is enhanced by more efficient data handling in the hospitals, which has become possible through the digitization of diagnostic information. The inevitable integration of medical data also offers other new possibilities, such as the use of pre-operatively acquired images during surgical procedures.

This chapter presents the principles of nuclear imaging methods and some cases studies and future trends of nuclear imaging. It discusses too the recent developments in image analysis and the possible impact of some important current technological progression on nuclear medical imaging. The survey is limited to developments for hospitals, mainly within the product range of some famous and emerging international companies.

2. Principles of nuclear medical imaging and image analysis

In addition to conventional gamma scintigraphic imaging, the two major nuclear imaging techniques developed are Positron Emission Tomography (PET) and Single Photon Emission Computed Tomography (SCECT). Both imaging modalities are now standard in the major nuclear medicine services.

2.1. The conventional scintigraphic imaging

2.1.1. The Anger gamma camera

The principle of radiation detection is based on the interaction of these radiations with the matter. When a gamma photon enters in interaction with a detector material, it loses its energy mainly in the form of ionizations or excitations. The excited atoms return to their ground state through the emission of secondary low energy gamma photons. The incident gamma photon can be partially or totally absorbed (photoelectric effect). In the first case, the energy loss is accompanied by a deviation of the photon (Compton scattering). The photon loses "memory" of its initial place of issue. So the photoelectric effect is the right phenomenon which must be considered when we interest to the gamma-ray emission site.

In the gamma camera, the detection medium is historically a NaI scintillation crystal typically doped with thallium. This crystal is able to emit light especially through a fluorescence process after the excitation of its molecules by a charged particle (electron). The density of NaI is 3.67 g/cm3 and its atomic number 50. Its time of scintillation (fluorescence) is 230 nm and the maximum light emission is at 4150 Angstroms wave length. Its refractive index is 1.85, and it is relatively transparent to its own light; about 30% of emitted light is transmitted to the detection chain [1]. The energy resolution can reach 7-8% at 1 MeV and the constant time of their pulse is equal to ~10^{-7} sec. The detection efficiency of NaI is quite large, of the order of 40 photons/keV. Indeed, gamma-ray energy of 100 keV transferring all its energy in the crystal results in the creation of approximately 4000 fluorescence light photons. These photons are collected by the photocathode of a photomultiplier tube (Figure 1).

For the detection of the secondary light photons generated in the crystal by the interaction with the incident gamma radiations, a photomultiplier tube (PMT) located behind the scintillator is used (Figure 1). At the level of the PMT photocathode, each light photon is converted to electrons. These electrons are then accelerated and multiplied by ten dynodes polarized by a gradually increasing voltage, and finally collected by an anode placed at the other side of the PMT where they give birth to an electrical impulse. This pulse has an amplitude proportional to the energy of the detected gamma-ray.

The output signal is amplified by the PMT. Its amplitude is measured, digitized and stored. Numerical analysis enables to obtain a spectrum (number of photons detected as a function of

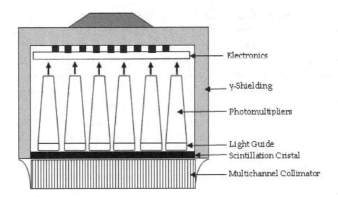

Figure 1. Main components of Gamma-camera.

their energy) characteristic of the detected gamma-rays. Detection time (acquisition) should be sufficient to obtain good counting statistics. The theoretical gamma-rays spectrum reaching the crystal is a line spectrum; the spectrum is continuous (Figure 2). The spectrum includes the total energy peak corresponding to gamma directly emitted by the radioactive source without any interaction before reaching the crystal and a background of lower energies due to the partial absorption of gamma by Compton scattering. Compton scattering in the path of the photon is changed making it impossible to locate its transmitter site. It is therefore necessary to take into account only the events corresponding to the photoelectric interactions at the level of the crystal with the total emission energy. This is achieved by the intermediate of a "window" for selecting the double-threshold energy (pulse height analyzer).

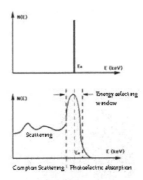

Figure 2. Gamma-rays spectrum at the level of the crystal detector (ideal (top) and real (bottom) cases).

The width of the peak of total absorption depends essentially of the random statistical fluctuations of the gain of the PMT. The width at half maximum ΔE relative to an average energy E_0 defines the energy resolution $\Delta E/E0$. The energy resolution of PMT is about 10% at 140 keV (emission peak of technetium-99m). The pulses selected by the pulse analyzer (maximum intensity) are directed to a time scaling circuit having a time integrator which then delivers a count rate in counts per second (cps). This count rate can be correlated to the real activity of the source after a number of corrections taking into account in particular the geometric efficiency and the detection performance of the detection chain. For very high source activity, the detector response is no longer linear so that a number of events are not taken into account. The lapse of time in which these events are lost (not counted by the detector) is called the dead time. In practice, it is usual, to work under conditions such that the detection dead time correction is not necessary (medium activity source).

The Anger gamma scintillation camera (Figure 3) uses the information provided by the amplitude of the electrical pulse not only to measure the energy of the detected radiation, but also to locate in the space the emission site of this radiation.

The camera developed by Anger in 1953 has a crystal of sodium iodide (NaI) thallium activated. It can take single crystal of large dimensions, up to 60x50 cm2 with a thickness ranging from 1/4 inch to 1 inch [1]. These crystals are fragile and are highly sensitive to shocks and moisture. The surface of the crystal is covered with a large number of PMTs (between 50 and 100). When scintillation occurs, the sum of the output signals of all the MPTs provides the energy lost in the volume of the scintillator (Z coordinate). The large number of PMTs ensures the collection of maximum light. Moreover, the amplitude of the output signal of PMT varies with the distance between the centre of the photocathode and the place where the scintilaltion is produced is in the crystal. The amplitude distribution of the output pulses of the PMT then provides the location information (X and Y coordinates) by means of a computer listing. For each photon interacting with the detector is thus obtained location coordinates (X and Y) and a value of the energy given or lost in the crystal (Z coordinate). An amplitude analysis allows selecting only the photon energy characteristic of the radionuclide used (eg. 140 keV for 99mTc) having lost all their energy in the crystal (photoelectric peak).

Figure 3. Gamma-camera called also Anger camera.

The scintillation Gamma-camera was used originally for planer projection imaging is mainly composed by the following components:

2.1.1.1. The collimator

The scintigraphic image corresponds to the projection of the distribution of radioactivity on the crystal detector. Gamma rays cannot be focused using lenses as in the case of light. The use of a special kind of collimator can permit just to one direction gamma rays to reach the crystal, the most common being perpendicular to the crystal. A collimator is a wafer usually lead wherein cylindrical or conical holes are drilled along a system axes determined. Gamma-ray where the path does not borrow these directions is absorbed by the collimator before reaching the crystal. The partition (wall) separating two adjacent holes i called "septa". The thickness of lead is calculated to cause an attenuation of at least 95% of the energy of the photons passing through the septa. The most commonly used collimator is the parallel holes. It retains the dimensions of the image. For non-parallel collimators, the dimensions of the image depend on the geometrical disposition and the divergence or convergence nature of the collimator. This leads to a geometric distortion must be taken into account. The efficiency of a collimator is the fraction of radiation passing through the collimator (without any interaction), reaching the crystal and effectively participating in the image formation. The collimator resolution corresponds to the accuracy of the image formed in the detector. Resolution improves with increasing thickness of the septa at the expense of collimator efficiency. A good compromise is to find the realization of a collimator performance depends on the intrinsic characteristics of the detector and the use we want to make [2].

2.1.1.2. The scintillator crystal

The γ-camera crystals are generally composed of NaI(Tl). Features that make this crystal desirable include high mass density and atomic number (Z), thereby effectively stopping γ photons, and high efficiency of light output [3, 4]. The most important characteristics of the crystal that must be ensured are: 1) high detection efficiency, 2) high energy resolution, 3). low decay constant time and a light refraction index close to the glass one. Most current cameras incorporate large (50 cm×60 cm) rectangular detectors. While expensive, the larger field of view results in increased efficiency. In early designs, crystals were often 0.5 inches thick, which was well-suited for high energy γ photons. In more recent implementations of the γ-camera, crystals only 3/8-inch or 1/4-inch thick are used, which is more than adequate for stopping the predominantly low-energy photons in common use today and which also results in superior intrinsic spatial resolution.

2.1.1.3. The photomultipliers tubes

Their role is to convert light energy emitted by the crystal to an electrical signal that can be exploited in electronic circuits [3, 5]. This is achieved by the combination of several elements, placed in a vacuum to allow the flow of electrons. The first element, placed in contact with the crystal is the photocathode, metal foil on which the light photons are able to extract electrons. These electrons are attracted to the first dynode by the application of a high voltage between it (positively charged) and the photocathode. The electrons acceleration allows them to extract a much larger number of electrons from the dynode. Then there are several cascading dynodes, on which the same phenomenon is repeated. The successive dynodes are submitted to potentials higher and higher. From a dynode to another, we obtain a cascade of electrons more intense (amplification phenomenon), which ultimately results in a measurable electric current. This current is collected by the last element called anode and a real electrical signal is generated (Figure 4).

Figure 4. PMTs disposition in a Gamma-camera. Generally a hexagonal shape of PTM is preferred then a circular because it well cover the detection area. Additional very small PMT can also be used between principal PMT for best detection area covering (CEM, Rennes, France).

2.1.2. Gamma scintigraphic imaging

Scintigraphy is a method designed to reproduce the shape or to measure the activity of an organ by administering a product which contains an element which emits radioactivity, an isotope. The radioactivity emitted by the isotope is picked up by special detectors called gamma-cameras counters described above. Generally, the dose is administered to a patient in need of scintigraphy is safe for the body (except for pregnancy). The data acquisition principle is illustrated on the diagram of Figure 5.

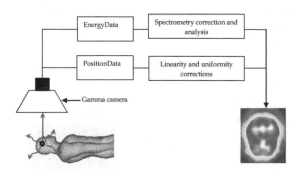

Figure 5. Illustration of data acquisition in planer gamma scintigraphy.

The use of radioactive tracers that are introduced in the living system to study its metabolism dates from 1923 when de Hevesy and Paneth studied the transport of radioactive lead in plants [6]. In 1935, de Hevesy and Chiewitz were the first to apply the method to the study of the distribution of a radiotracer (P-32) in rats [7]. The major development of scintigraphic imaging started with the invention of the gamma camera by Anger in 1956 [1]. In parallel, positron imaging was developed. Both imaging modalities are now standard in the major nuclear medicine departments.

The tracer principle, which forms the basis of nuclear imaging, is the following: a radioactive biologically active substance is chosen in such a way that its spatial and temporal distribution in the body reflects a particular body function or metabolism. In order to study the distribution without disturbing the body function, only traces of the substance are administered to the patient [8, 9].

The radiotracer decays by emitting gamma rays or positrons (followed by annihilation gamma rays).The distribution of the radioactive tracer is inferred from the detected gamma rays and mapped as a function of time and/or space.

The most often used radio-nuclides are Tc-99m in 'single photon' imaging and F-18 in 'positron' imaging. Tc-99m is the decay daughter of Mo-99 which itself is a fission product of U. The half-life of Tc-99m is 6h, which is optimal for most metabolic studies but too short to allow for long time storage. Mo-99 has a half-life of 65h. This allows a Mo-99 generator (a 'cow') to be stored and Tc-99m to be 'milked' when required. Tc-99m decays to Tc-99 by emitting a gamma ray with an energy output of 140 keV. This energy is optimal for detection by scintillator detectors. Tc-99 itself has a half-life of 211100 years and is therefore a negligible burden to the patient [8, 9].

F-18 is cyclotron produced and has a half-life of 110 minutes. It decays to stable O-18 by emitting a positron. The positron loses its kinetic energy through Coulomb interactions with surrounding nuclei. When it is nearly at rest, which in tissue occurs after an average range of less than 1 mm, the probability of a collision with an electron greatly increases and becomes one. During the collision matter-antimatter annihilation occurs in which the rest mass of the electron and the positron is transformed into two gamma rays of 511 keV. The two gamma rays originate at exactly the same time (they are "coincident") and leave the point of collision in almost opposite directions [9].

Different modalities of scintigraphic acquisition are possible:

1. Static acquisition with a detector in a fixed position relative the patient: examination of thyroid, kidney....

2. Scanning of the whole body: succession of static images joined: the detector move simultaneously and scan the patient's body from head to foot. The bone scan is a routine application.

3. Tomographic acquisition: The Positron Emission Single Photon (SPECT): detectors rotate around the patient to obtain in a digital representation of a 3D radioactive distribution of the body: chest, pelvis, skull....

4. Dynamic acquisition as a function of time: a number of successive static images used to reconstruct a video to study some interesting dynamic biological processes. Interesting applications are: kidney and bone phase's vascular scans and scintigraphy of the heart ventricle.

5. ECG[1] gated acquisition: used for tomographic myocardial scintigraphy. In this application, detectors are arranged in the shape of an "L» and simultaneously record the radioactivity from the myocardium and the electrical activity of the heart. Thus it is a dynamic acquisition synchronized by the heartbeat which is recorded by ECG.

2.2. Single photon emission computed tomography

This medical imaging method was introduced in 1963 by Kuhl and Edwards [10]. Known by the acronym SPECT (Single Photon Emission Computerized Tomography), this imaging method is equivalent in scintigraphy to Computed Tomography (CT) in radiology. The injected radioactive tracers emit during their disintegration gamma photons which are detected by an external detector, after passing through the surrounding tissue. Because the gamma photons emission is isotropic, a collimator is placed before the detector to select the direction of the photons to be detected. Thus, if we call $f(x, y, z)$ the distribution of radioactivity emitted point $\{x, y, z\}$ per unit solid angle, the number of photons detected at the point $\{x',y'\}$ of the detector is equal to (Figure 6) [11]:

$$N(x',y') = \int_L f(x,y,z)ds \tag{1}$$

Where L is the line given by the direction of the channel's collimator and passing through the point (x',y'). As in CT, the various projections are obtained by rotating the detector around the object (patient).

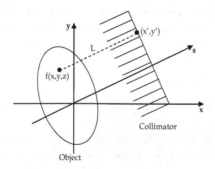

Figure 6. Detection principle in SPECT imaging.

---

1 ECG : Electrocardiogram.

In SPECT, the main radioactive isotopes are technetium-99m, Iodine and Thallium-201, which is used primarily for studies on the heart. At the opposite of PET system, the collimator is an indispensable component in a SPECT machine. The first collimators used were two-dimensional parallel channels (Figure 7, a). By rotating the detector & collimator assembly around the patient, two-dimensional projections are obtained, and the distribution of radioactivity may be 3D reconstructed slice by slice. These parallel collimators are used in the vast majority of SPECT systems used in Nuclear Medicine services. The resolution of these systems varied from 10 to 15 mm.

To increase the sensitivity and resolution of SPECT systems, converging channels collimators were developed (Figure 7, b). The first proposed included a series of converging channels to a focal line which is parallel to the rotation axis of the system [12]. This system is therefore equivalent to a scanner used in X-ray fan beam tomography where 3D image is reconstructed slice by slice. For imaging small organs such as heart and brain, a converging cone collimators is used [13, 14]. This last collimator allows obtaining magnification of the object in all directions (cross and longitudinal). This kind of collimators can be used only for small field tests, so for small structures, the size of the detectors has not increased. With these systems, image data registration is completely 3D as well as in cone beam X-ray tomography, and therefore reconstruction is not performed slice by slice. In these systems, it is important to be able to shift the head of the detector relatively to the rotation axis, thereby to perform trajectories other than circular. In addition to the fact that this shift allow to complete the set of projections, such a shift is interesting to avoid obstacles, such as shoulders brain imaging. Finally, other kinds of collimators are also available for SPECT such as diverging and pinhole collimators. Diverging collimator (Figure 7, c) is reserved to large structure imaging. Pin-hole collimator (Figure 7, d) allows obtaining a mirror image with a variable magnification function of collimator depth and object to collimator distance. This collimator is suitable for small structures imaging such as thyroid and hip.

2.3. Positron emission tomography

Positron emission tomography (PET) is a medical imaging modality that measures the three-dimensional distribution of a molecule labelled with a positron emitter. The acquisition is carried out by a set of detectors arranged around the patient. The detectors consist of a scintillator which is selected according to many properties, to improve the efficiency and the signal on noise. The coincidence circuit measures the two 511 keV gamma photons emitted in opposite directions resulting from the annihilation of the positron. The sections were reconstructed by algorithms, the same but more complex than those used for conventional CT, to accommodate the three-dimensional acquisition geometries. Correction by considering the physical phenomena provides an image representative of the distribution of the tracer. In PET scan an effective dose of the order of 8 mSv is delivered to the patient. This technique is in permanent evolution, both from the point of view of the detector and that of the used image reconstruction algorithms. A new generation of hybrid scanner "PET-CT" provides additional information for correcting the attenuation, localize lesions and to optimize therapeutic procedures. All these developments make one PET fully operational tool that has its place in medical imaging.

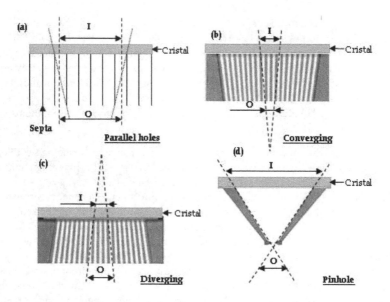

Figure 7. Different kinds of collimators used with SPECT imaging system (O: object, I: image).

Positron emitters are radioactive isotopes (^{11}C, ^{13}N, ^{15}O, ^{18}F) which can easily be incorporated molecules without altering their biological properties [15-22]. The first ^{18}F labelled molecules were synthesized to late 1970s. At the same time, were built the first emission tomography scanners (PET cameras) used in a clinical setting. Since the 1970, many studies conducted by research centres and industrialists have allowed the development of PET to perform tests whole body, in conditions of resolution and adapted sensitivity. Until the last decade, PET was available only in centres equipped with a cyclotron capable of producing the different isotopes. However, today's growing role PET in oncology is reflected in the rapid spread of this medical imaging modality in hospitals. The operation of these structures is based on the installation of PET machine, and the implementation a network distribution radio-pharmaceutical marked by ^{18}F, characterized by a half life of 110 minutes. The most widely used molecule is the Fluorodeoxyglucose (FDG) labelled with fluorine 18 (^{18}F-FDG) due to its many properties and advantages. Generally to find the right tracer molecule, a close look into the designated processes and the related biochemistry is necessary, the following gives a short overview:

- Metabolism and general biochemical function;

- Receptor-ligand biochemistry;

- Enzyme function and inhibition;

- Immune reaction and response;

- Pharmaceutical effects.

• Toxicology (carcinogen and mutagenic substances).

The realization of a PET scan is the result of a set of operations, since the production of the isotope, the synthesis of the molecule, the injection of the radioactive tracer, the detection of radiation, the tomographic reconstruction, and finally the application of a series of corrections to provide image representative of the distribution of the tracer within the patient.

The main physical characteristics of isotopes used in PET are summarized in Table 1.

Isotopes	^{11}C	^{15}N	^{15}O	^{18}F	^{76}Br
Maximum kinetic energy of β⁺ (MeV)	0.98	1.19	1.72	0.63	3.98
Period (mn)	20.4	10.0	2.1	109.8	972
Maximum Free path in water (mm)	3,9	5	7,9	2,3	20

Table 1. Physical characteristics of the main isotopes positron emitters used in positron emission tomography (PET).

The principle of PET is based coincidence 511 keV Gamma-photons detection (created by positron annihilation) by considering the parallelepiped joining any two detector elements as a volume of response (Figure 8, a). In the absence of physical effects such as attenuation, scattered and accidental coincidences, detector efficiency variations, or count-rate dependent effects, the total number of coincidence events detected will be proportional to the total amount of tracer contained in the tube or volume of response. Both Two and three dimensional modalities are available for one scan and it depends on the collimator-Detector system used. In two dimensional PET imaging, only lines of response lying within a specified imaging plane are considered (Figure 8, b). The lines of response are then organized into sets of projections. The collection of all projections obtained by rotation around the patient forms a two dimensional function called a sonogram which will be used for 2D image reconstruction. Multiple 2-D planes are can be stacked to form a 3-D volume. In fully three-dimensional PET imaging, the acquisition is performed both in the direct planes as well as the line-integral data lying on 'oblique' imaging planes that cross the direct planes, as shown in figure 8 c. PET scanners operating in fully 3-D mode increase sensitivity, and thus reduce the statistical noise associated with photon counting and improve the signal-to-noise ratio in the reconstructed image.

2.4. PET and SPECT images processing and analysis

Tomographic slices are reconstructed from the acquired projection data using either analytic or iterative algorithms. Analytic reconstructions represent an exact mathematic solution, and there is a general solution for true projection data: filtered backprojection. Although filtered backprojection is a relatively efficient operation, it does not always perform well on noisy projections and, as is the case with SPECT and PET data, it generates artifacts when the projections are not line integrals of the internal activity. Iterative algorithms are a preferred alternate method for performing SPECT reconstruction, and over the past 10 years there has been a shift from filtered backprojection to iterative reconstruction in most clinics [23-26]. The

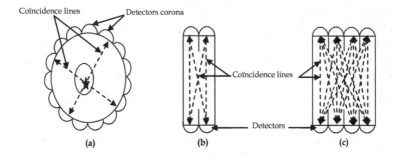

Figure 8. Principle of PET imaging and 2D and full 3D image acquisition modes.

big advantage of the iterative approach is that accurate corrections can be made for all physical properties of the imaging system and the transport of γ-rays that can be mathematically modeled. This includes attenuation, scatter, septal penetration in the case of SPECT, and spatial resolution. In addition, streak artifacts common to filtered backprojection are largely eliminated with iterative algorithms. A major advance was the introduction of the ordered-subset expectation maximization approach, which produces usable results with a small number of iterations.

In each study, the PET or SPECT images selected for statistical analysis are registered, smoothed and intensity normalized and this because of the following objectives:

• Registration is required to align the data sets, which is an important step for any kind of voxel-by-voxel-based image analysis.

• Smoothing effectively reduces differences in the data, which cannot be compensated for by registration alone, such as intrapatient variations in pathology, and the resolution of the reconstruction of scans. Another reason for smoothing is the reduction of noise.

• Intensity values of the data sets may vary significantly, depending on the individual physiology of the patient (e.g., injected dose, body mass, washout rate, metabolic rate). These factors are not relevant in the study of the disease, and need to be eliminated using intensity normalization, to obtain meaningful statistical comparisons during multivariate analysis.

Key PET and SPECT image processing parameters include also the following:

1. Filtering: improve image quality by removing noise and blur;

2. Reconstruction: by analytical or iterative methods;

3. Motion correction: recommended to reduce motion blur due to object motion;

4. Attenuation correction: identifying source of attenuation for image correction;

5. Quantification: assessment by image quantification of the affected area;

6. Normal database: reference used for calculation of extent and severity of defect;

7. Segmentation: process of partitioning a digital image into multiple segments to simplify and/or change the representation of an image into something that is more meaningful and easier to analyze;

8. Volume fraction calculation.

In addition to these pre-processing methods which have an impact on the interpretation of the results, there are other processing methods that must be applied to SPECT image to extract essential information according to the studied pathologic case. Thus, SPECT images can be processed by various methods such as: 1) "Principal Components Analysis (PCA) which is a multivariate analysis method that aims at revealing the trends in the data by representing the data in a dimensionally lower space[27], 2) "Discrimination Analysis (DA)" used to identify a discrimination vector such that projecting each data set onto this vector provides the best possible separation between population groups subject to SPECT study and 3) Bootstrap Resampling which is applied to evaluate the robustness and the predictive accuracy of the PCA and DA approach [28].

3. Recent development in nuclear imaging and image analysis

3.1. Recent advances in SPECT and PET imaging systems

The key technology in the development of SPECT and PET systems for static or dynamic image acquisition is embodied in the development of the detector, or rather, the detector chain. Although it has already reached a high degree of perfection, continuous improvements are still increasing the performance of, for example, the scintillator material, which is a critical component in the chain. The time of flight camera, introduced by Philips Medical Systems in the 1980s, is replacing the conventional Anger camera and offers significant improvements in image quality. The trend here is towards higher resolution where, for certain applications, 2048 x 2048 pixel matrices will be used. In addition to continuous improvements in the detector chain, there are also radically novel approaches which dispense with the need for a semiconductor detector. A detector based on scintillator crystals coupled to hybrid photodetectors that provides full 3D reconstruction in PET imaging with high resolution and avoiding parallax errors developed during last ten years are actually available [29, 30].

Another improvement is SPECT systems provision on a single stand of rotation of several (two or three) detecting heads, allowing examination time reduction and detection sensitivity increasing. In addition, one of the heads can record a transmission coefficient image induced by a radioactive external gamma source photons of the same energy as those issued by the tracer during the examination. These acquisitions are then used to correct the effect of self-absorption.

Development of SPECT and PET systems much more efficient enable major advances in the clinical use of these techniques with very widespread applications field. Additional development may include research on more efficient scintillators, the use of more adequate recording geometries, such as the conical geometry for example, accompanied sure with the development of robust reconstruction algorithms.

Time-of-Flight technology has always held the promise of better PET imaging. Philips delivered on that promise with its innovative Astonish TF technology. Now with 4D TOF, Philips continues to push the envelope of PET imaging performance. See how 4D TOF Innovation is making an impact on PET imaging.

Design of Hybrid machines has been a very interesting research and technologic development axe in nuclear imaging during last fifteen years. Indeed, many hybrid PET-CT, SPECT-CT and PET-MRI machines were manufactured offering a variety of very interesting diagnostic applications by the combination of results of two imaging methods allowing the revelation of a very interesting pathologic information that cannot be revealed by a single technique alone. PET-CT is creating a new benchmark in imaging and analysis of cardiovascular disease. PET-CT enables the combination of PET myocardial perfusion and viability imaging with CT coronary angiography and calcium scoring in a single integrated environment. In oncology, it provides the integration of metabolic data from PET and anatomical data from CT.

SPECT-CT is a system designed entirely for nuclear medicine and has particular value in the cardiology cycle of care. This hybrid machine allows table to remain stationary in many cases, eliminating complexities inherent in table indexing, acquires the entire heart volume in just one rotation and permits patients to breathe normally during SPECT and CT acquisitions. In oncology, it plays an important role in diagnosis, treatment, and follow-up in the oncology cycle of care, including the use of low-dose localization and aids better visualization that is especially valuable during studies and in bone imaging.

Researchers continue to develop new ways of using PET. One recent development has been the combination of PET and MRI[2] into a single apparatus. Compared to CT, MRI generally provides more detailed images, which can aid in the more precise localization of cancerous growths. A hybrid PET-MRI scanner simultaneously delivers functional information plus anatomy and tissue characterization (soft tissue contrast and blood vessel physiology), from a state-of-the-art MRI scanner. At the same time, it provides metabolic imaging from PET technology. Fusing these images gives the best of both worlds, providing greatly superior information to what you'd get from either machine individually

Actually, the main hybrid machines routinely used in hospitals are the following:

3.1.1. PET-CT

The first machine was created by University of Pittsburgh physicist David Townsend and engineer Ronald Nutt; the PET-CT machine was called the "Medical Science Invention of the Year" by *Time* magazine in 2000. After giving entire satisfactory at the research tests level and their importance in oncology and cardiology were well demonstrated, many international companies were interested in the fabrication of such kind of hybrid imaging machine. Actually, the market is shared mainly between General Electric (GE), Philips and Siemens (Figure 9).

---

2 MRI: MagneticResonance Imaging.

GE offers a variation in its range of PET-CT "Discovery ST" machine to meet the specific clinical needs. After the Discovery ST oriented oncology and cardiology, the GE Discovery VCT sells dedicated cardiology is associated with a 64-slice scanner. The latest version offers a higher spatial resolution responding to neurological applications. GE ST machines are available in versions scanner 4, 8 or 16 cups. The 2D acquisition abandoned by other manufacturers is optional and defended by GE to obtain less noisy images (useful for some advanced applications or for overweight patients) and for new applications mostly outside the scope FDG. GE believes that the increase of activity of PET-CT will be around 50% in the next three years and examines the association of PET and MRI modalities. The contribution of MRI compared to CT is questionable, except perhaps in functional imaging.

PHILIPS GEMINI PET/CT scanners combine the Brilliance CT technology, that is well-suited to cardiac imaging with its wide-coverage submillimeter imaging, ultra fast acquisition times and Rate Responsive image acquisition technology that adapts to the patient's heart rate and rhythm during acquisition. GEMINI PET/CT scanners deliver high spatial resolution and high sensitivity PET imaging resulting in improved image quality when imaging the short-lived radiopharmaceuticals used with cardiac PET. Philips PET-CT hybrid machines ALLEGRO maintain in the range GEMINI.

SIEMENS works to upgrade the install PET-CT around the world. The range of PET-CT, BIOGRAPH marketed since 2000 continues to benefit from developments. After improving the sensitivity BGO crystals by replacing the LSO crystals, SIEMENS in 2004 increased the detection speed by introducing a new channel detection (PICO 3D) with the coincidence window is only 4.5 ns and improved spatial resolution due to detector Hi-Rez (block 13 x 13 x 8 against 8 elements far). Note that BIOGRAPH have a tunnel of 70 cm diameter field used in whole to acquire PET scanner. This criterion is important for obese patients.

Discovery
GE Healthcare

Gemini
Philips Medical Systems

Biograph
Siemens Medical Solutions

Figure 9. Example of commercially PET-CT scanners.

3.1.2. SPECT-CT

A variety of SPECT-CT scanners are nowadays available in many hospitals and oncology centres (Figure 10). GE proposes a robust SPECT-CT hybrid machine called "Infinia" which is a dual-head, large field for general applications. The Infinia has an open stand. It is available with SPECT thick crystals (5/8th) or thin (3/8th) depending on the intended application. It is available in solo or in combination with a scanner. The Infinia Hawkeye 4 SPECT/CT from GE

Healthcare builds upon its performance with a wealth of innovations, from enabling low dosage and improved acquisition times to enhancing imaging results through scatter correction modeling and reduction, motion detection and correction, and accurate attenuation correction. Hawkeye 4 should respond to all applications except exams angio CT or cardiology.

PHILIPS approaches the market hybrid machines by combining existing methods in its range. The hybrid machine called PRECEDENCE. Precedence SPECT/CT system offers the combination of functional data from SPECT with high-resolution anatomical detail from a multi-slice diagnostic CT scanner to give clinicians a new standard of diagnostic confidence.

When SPECT functional data is fused with CT, the location and extent of disease may be better visualized and treated.

SEIMENS "Symbia" SPECT-CT hybrid machine is integrated SPECT and diagnostic multislice-CT bring a whole new dimension to nuclear medicine. With the ability to provide precise localization of tumors and other pathologies before disease reveals itself, Symbia has the potential to revolutionize treatment planning for cancer, heart disease, and neurological disorders. Symbia has enormous potential for cardiac imaging, revealing even the hard-to-detect conditions that carry the highest risk for patients.

The GE Infinia   The Philips Precedence   The Siemens Symbia
Hawkeye 4 SPECT/CT scanner   SPECT/CT scanner   SPECT/CT scanner

Figure 10. Examples of SPECT-CT hybrid scanners.

3.1.3. PET-MRI

Simultaneous PET and MRI scans eliminate the need to move patients from one imaging unit to another, making it easier to combine data from both scans to produce enhanced details. The scanner also exposes patients to significantly lower radiation levels than an older combined scanning technique, PET-computed tomography (CT). PET-MRI scanner is used in understanding certain types of malignancies, such as cancers of the brain, neck and pelvis because the anatomy is very complex in those areas, and combined PET-MRI should produce a more detailed reading of the intricate boundaries between disease and healthy tissue. The integration of PET and MRI for simultaneous scanning was a complex task because powerful MRI magnets interfered with the imaging detectors on the PET scanner. But scientists overcome this problem and PET-MRI scanners are nowadays available for research and patient care (Figure 11).

In 2010, Philips unveiled its own solution which involves a 3T MR and a high resolution PET scanner with an integrated rotating table that passes the patient from one machine immediately into the other. Philips Ingenuity TF PET/MR is a new modality so original and resourceful that it offers Astonish Time-of-Flight technology combined with the superior soft tissue imaging of Achieva 3.0T MRI in a whole-body footprint.

In 2011, Siemens Healthcare said that its hybrid PET-MRI scanner received USA Food and Drug Administration clearance. The device, the Biograph mMR, is the first integrated PET-MR device capable of doing simultaneous whole-body magnetic resonance imaging and positron emission tomography scans. It combines a 3-Tesla MR system with PET detectors, giving doctors the morphological and soft tissue information from MR with the cellular and metabolic activity data from PET.

Philips TF PET/MRI Combo   SEIMENS Biograph PET/MRI Scanner

Figure 11. Actually available PET-MRI hybrid scanners.

3.2. Recent developments in nuclear medical image acquisition and analysis

In addition to conventional nuclear image processing methods described above, Registration and Validation are also a very important research axes in nuclear imaging. In this section, we present the state-of-the-art and research topics regarding only these two axes.

3.2.1. Registration

There is increasing interest in being able to automatically register medical images from either the same or different modalities. Registered images are proving useful in a range of applications, not only providing more correlative information to aid in diagnosis, but also assisting with the planning and monitoring of therapy, both surgery and radiotherapy. The classification of registration methods is classically based on the criteria formulated by van den Elsen, Pol & and Viergever [31]. Many basic criteria can be used, which each can be developed and subdivided again [32, 33]. The main are the following:

1. Dimensionality: 2D or 3D only spatial dimensions or time series with spatial dimensions;

2. Nature of registration basis: Extrinsic, Intrinsic or Non-image based (calibrated coordinate systems);

3. Nature of transformation: rigid, affine, projective, or curved;

4. Doman of transformation: local, global or interaction;

5. Interaction: interactive, semi-automatic or automatic;

6. Optimization procedure: parameters computed or parameters searched for;

7. Modalities involved: mono-modal, multi-modals, modality to model or patient to modality;

8. Subject: intrasubject; intersubject or atlas;

9. Object: head, abdomen, limbs, thorax...

Although great advances have been made in basic nuclear medicine imaging in both the detection and estimation tasks, personalized medicine is a challenging goal. It requires the ability to detect many different signals that are specific to a patient's disease. That requirement has led to the increasing development of hybrid imaging systems.

The development of image reconstruction algorithms, simulation tools, and techniques for kinetic model analysis plays an important role in the right interpretation of the generated image signals. Development of these software tools is essential to accurately model the data and thereby quantify the radiotracer uptake in nuclear medicine studies. The ability to perform this task in practice has benefited from the increased availability of powerful computing resources. For example, an iterative image reconstruction algorithm with data corrections built into the system model was considered to be impractical a decade ago. Yet, this type of algorithm can now be used to generate images in a practical amount of time in both the research laboratory and the clinic Leaders in instrumentation and computational development in nuclear medicine from universities, national laboratories, and industry were solicited for commentary and analysis.

3.2.2. *Validation*

The ability of nuclear imaging devices to provide anatomical images and physiological information has provided unparalleled opportunities for biomedical and clinical research, and has the potential for important improvements in the diagnosis and treatment of a wide range of diseases. However, all nuclear imaging devices suffer from various limitations that can restrict their general applicability. Some major limitations are sensitivity, spatial resolution, temporal resolution, and ease of interpretation of data. To overcome these limitations, scientists have worked particularly on: on: 1) Development of technological and methodological advances that improve the sensitivity, spatial resolution and temporal resolution, 2) Development of multi-modality approaches that combine two (or more) biomedical imaging techniques. In addition to these two research areas, validation of nuclear imaging technologies and methodologies is uncontainable to develop nuclear imaging and medicine. Development of "multi-modality" approaches could be used to combine information that might not be available from a single imaging technique or to compare and validate results obtained with one imaging technique with results obtained using another imaging technique. Thus, development and improving approaches for analysis and optimization of complex multi-component biomedical imaging devices is highly required. The validation methods are classified in the following main categories:

1. Statistical validation methods;

2. Validation with phantoms;

3. Clinical validation.

To date there is very little in terms of validation and standardizing the validation process in nuclear image processing. Further research is needed in validation for nuclear image-processing as issues concerning validation are numerous. Clinically relevant validation criteria need to be developed. Mathematical and statistical tools are required for quantitative evaluation or for estimating performances in the absence of a suitable reference standard. The diversity of problems and approaches in medical imaging contributes significantly to this. Validation data sets with available accuracy reference are required. Comprehension of clinical issues and establishment of robust therapy protocols is also required. Indeed, validation is by itself a research topic where methodological innovation and research are required [34].

4. Cases studies and future trends of nuclear imaging

Current clinical applications of nuclear medicine include the ability to:

*   diagnose diseases such as cancer, neurological disorders (e.g., Alzheimer's and Parkinson's diseases), and cardiovascular disease in their initial stages through use of imaging devices including PET-MRI, PET-CT and SPECT-CT;

*   provide molecularly targeted treatment of cancer, and certain endocrine disorders (including thyroid disease and neuroendocrine tumors);

*   Non-invasively assess a patient's response to therapies, reducing the patient's exposure to the toxicity of ineffective treatments, and allowing alternative treatments to be started earlier.

The use of nuclear hybrid imaging, particularly PET-CT, is expanding rapidly. More recently, positron emission tomography (PET) has increased its applications in total body imaging to include the postoperative orthopedic patient. PET and PET-CT scanning for postoperative infection has also been investigated in the spine, also showing good results, with increased specificity for infection in contrast to routine three-phase bone scan or combination radiotracers [35]. The increasing specificity of nuclear medicine agents continues to broaden nuclear medicine applications in the postoperative musculoskeletal imaging setting.

The development of SPECT and SPECT-CT is a logical consequence of the previous success of PET-CT, the first of these hybrid imaging techniques. The introduction of this technique, about 10 years ago, meant a final advanced nuclear medicine in the field of oncology. Pushed forward by the scientific and commercial success of these PET-CT, the industry developed the SPECT-

CT, a technology similar to the exams conventional (= non-PET) nuclear medicine. Here too, the SPECT functional information is supplemented by information from CT coupled thereto. Within a single examination, SPECT-CT is able give the correct diagnosis of bone lesion corresponded to metastatic disease. In a general hospital, the SPECT-CT is also used in the development of pain syndromes of orthopedic or rheumatic origin, for example at the lumbar level ("back pain") or a knee. The success of SPECT-CT is that the bone scan shows osteoblastic lesions selectively cause pain and coupling with the CT image interpretation makes-SPECT abnormalities more accurately [36]. SPECT-CT is also successively used for the detection of sentinel lymph node scintigraphy. It allows the visualization of the effect or lymph vessels in which they lead and are thus likely to be the site of métastastiques cells. In principle (and in practice), if such individual nodes called "sentinel" are not found with the tumor cells, while cleaning, any additional node excision is unnecessary [36]. Among other undesirable side effects, thus avoiding impairment of lymphatic drainage of the upper limb and the onset postoperative thugs. SPECT-CT allows more accurate localization by this or these nodes but also give information on their volume, shape and density, all useful information for surgeons in their quest intraoperative these nodes. SPECT-CT in this area still has other potential applications, such as cancers of the prostate, cervix of the uterus and of the head and neck. Patients with thyroid cancer who develop recurrent disease is suspected are often subjected to whole body scintigraphic imaging after administration of a small activity of an isotope of iodine (iodine-123 or iodine-131). With SPECT-CT, better diagnosis of pulmonary embolism is also possible. Pulmonary embolism (PE) is indeed a common problem in cancer. Planar scintigraphic imaging of the normal, the diagnosis of PE is typically established by the demonstration of a mismatch, a defect of pulmonary perfusion with preserved ventilation, normal in the same territory. Here SPECT acquisitions of pulmonary ventilation (after inhalation aerosol technetium) and pulmonary perfusion (after injection of macro-aggregates of albumin technetium) will be combined with a CT scan of the lungs. The classically observed mismatches between ventilation (preserved) and perfusion (altered) will be confronted with anomalies of the CT scan in the corresponding regions [36]".

A review of applications of PET, PET-CT, SEPCT and SPECT-CT and their clinical benefits with an emphasis on oncologic applications is given below (Figures 12-18).

Figure 12. Thyroid scan with planar scintigraphy (99mTc04). Source: CEM, Rennes.

Rest (a) Effort   Rest (b) Effort

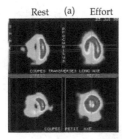 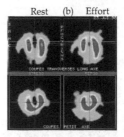

Figure 13. A SPECT slice of a patient's heart. SPECT is generally indicated for evaluation of coronary perfusion and myocardial viability. (a): showing anterior ischemia, (b): demonstration of a myocardial infarction. Source: CEM, Rennes.

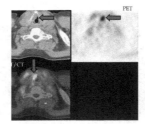

Figure 14. Larynx cancer demonstration and imaging with PET and CT images combination. Source: CEM, Rennes.

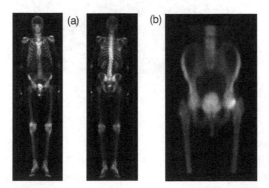

Figure 15. Bone SPECT Scan (Phosphonates -99m-Tc), (a): depicting bone metabolism in whole body: abnormal osteogenesis zones screening and surveillance (bone lesions carcinoma and other primary or metastatic bone lesions (Paget's disease, Osteomyelitis and fractures)), (b): SPECT bone scan showing left femoral neck fracture. Source: CEM, Rennes.

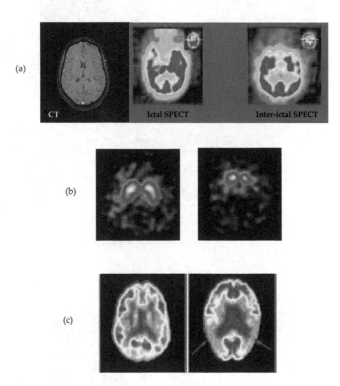

Figure 16. SPECT and PET applications in Neurology. These techniques are indicated in the diagnosis of Regional brain abnormalities (Cerebral perfusion) in and in vitro leukocyte marking (99mTc). (a) Epilepsy: SPECT can be very helpful in the localization of the epileptogenic zone and for mapping functional areas of the brain, such as those for language and motor function, (b) Parkinson: image from of a normal healthy case (left) and abnormal image in the case of early Parkinson's disease untreated, and (c) Alzheimer: PET scan of a normal volunteer (left) and a patient with Alzheimer's disease (right). Nuclear imaging devices help doctors diagnose such diseases in their initial stages. Sources: CEM, Rennes and Daniel Silverman, UCLA.

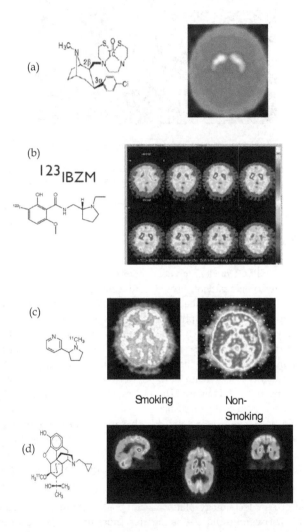

Figure 17. PET and SPECT neuro-receptors and neuro-transporters imaging with specific radio-marker molecules. (a): dopamine transporter, (b): dopamine receptor, (c): Nicotine receptor, and (d): Opioid receptor. Source: CEM, Rennes.

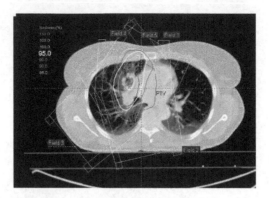

Figure 18. During radiotherapy planning FDG-PET-CT has been shown to be useful to better delineate the biologically active tumor volume and to distinguish between viable tumor tissue and non-specific changes due to previous surgical and/or radio therapeutic treatments. The figure present a planning for radiotherapy fields based on images from PET-CT in a patient with advanced stage lung carcinoma. Source: www.IAEA.org/…/gc54inf-3-att1_en.pdf.

5. Conclusions

In conclusion, PET and SPECT nuclear medical imaging have a clinical role in the evaluation of the postoperative oncologic patient, provided that the modalities are protocoled for the anticipated clinical concern and prescribed by the musculoskeletal physicians. Parameters and protocols include appropriate scintigraphic agent selection. These imaging techniques are also required to optimally visualize as much of the wide diversity of anatomical structures, and physiological and pathological processes, as possible. The success of nuclear imaging is due to the modality's ability to supply new clinical information which is useful for the routine care of large numbers of patients. The demand for more effective and less invasive therapy increases the need for real-time nuclear imaging. The choice of an imaging modality for a given procedure is determined by its ability to display both the patient's anatomy and the operator's instruments. Patient access and the safety of both patient and operator are also of major concern. Multi-modality (SPECT-CT, PET-CT and PET-MRI) imaging can often enhance medical decisions. Indeed, combining images from different origins in a workstation can facilitate this process to the benefit of the radiologist, referring physician and, ultimately, the patient.

The development of new technology platforms can contribute to accelerate, diversify, and lower the cost of discovering and validating new nuclear imaging probes, biomarkers, radiotracers, and labeled drugs, as well as new radiotherapeutic agents. The wide implementation of nuclear imaging techniques for local use in research and clinical programs requires the invention of new, small and low-cost miniaturized particle-accelerators and generators for producing short-lived radioisotopes. The invention of new detector technologies for PET and SPECT would contribute to enhance sensitivity as well as spatial and temporal resolution.

Finally, the development of new iterative algorithms and high-speed/high-capacity computational systems for rapid image reconstruction; would allow image data to be converted to quantitative parametric images pertaining to biological and pharmacological processes in disease.

Acknowledgements

I would like to thank Prof. Patrick Bourguet from the Department of Nuclear Imaging and Medicine, Centre Eugène Marquis (CEM), Rennes, France, for his support in the realization of this work and particularly for giving me the permission to use some examples of nuclear imaging applications and illustrations developed at his nuclear imaging laboratory.

Disclaimer

Data and statements expressed in this paper are those from the author and published bibliography cited in this work, and do not necessarily reflect organizations, laboratories and the firms which the author has mentioned as examples. The author does not endorse any equipment or material cited herein.

Author details

Faycal Kharfi*

Department of Physics, Faculty of Science, University of Ferhat Abbas-Sétif, Algeria

References

[1] Anger, H. O. Scintillation Camera. The Review of Scientific Instruments, 29(1); (1958). , 27-33.

[2] Anger, H. O. Scintillation camera with multichannel collimators. J Nucl Med, 5; (1964). , 515-531.

[3] Jonasson, T. Revival of a Gamma Camera, Master of Science Thesis (2003). Nuclear Physics Group, Physics Department, Royal Institute of Technology, Stockholm, TRI-TA-FYS 2003:40, 0028-0316X., 0280-316.

[4]   Zuckier, L. S. Principles of Nuclear Medicine Imaging Modalities in: Principles and Advanced Methods in Medical Imaging and Image Analysis. Dhawan, A.P., (ed), World scientific publishing; (2008). 109812705341

[5]   Photomultiplier tubes For Gamma Camera and Scintillation countinghttp:// sales.hamamatsu.com/assets/pdf/parts_R/Rpdfaccessed 20 January (2011).

[6]   Hevesy, G. de. J. Biochem(1923)., 439.

[7]   Hevesy, G. de. J. Chem Soc.; (1939)., 1213.

[8]   Buvat, I. Les principaux radiotraceurs et leurs applications. U494 INSERM, Paris; (2001).

[9]   Deconinck, F. ENS News, Issue Autumn; November (2006). (14)

[10] Kuhl, D. E, & Edwards, R. Q. Image separation radioisotope scanning Radiology 80; (1963)., 653-662.

[11] Peyrin, F. Introduction to 2D and 3D Tomographic Method Based on Straight Line Propagation : X-ray, Emission and Ultrasonic Tomography, Traitement du Signal-(1996)., 13-n

[12] Jaszczak, R. J, Chang, L. T, & Murphy, P. H. Single Photon emission computed tomography using multislice fan beam collimator. IEEE Tran Nucl Sci.,(1979). , NS-26, 610-618.

[13] Jaszczak, R. J, Floyd, C. E, Manglos, S. H, Greer, K. L, & Coleman, R. E. Cone beam collimation for single photon emission computed tomography: analysis, simulation, and image reconstruction using filtered backprojection. Med. Phys., (1986). , 13, 484-489.

[14] Gullberg, G. T, Zeng, G. L, Datz, F. L, Christian, P. E, Tung, C. H, & Morgan, H. T. Review of convergent beam tomography in single photon emission computed tomography. Phys. Med. Biol., N°3; (1992)., 37, 507-534.

[15] Qaim, S. M. Radiochim, Acta (2001). , 89, 223-232.

[16] Qaim, S. M. Radiochim, Acta (2001). , 89, 297-302.

[17] Bremer, K. H. In Ullmanns Enzyclopädie der technischen Chemie, Bd. 20: Anwendung von Radionukliden in der Medizin, VCH Weinheim; (1981). , 59-64.

[18] Wüstenberg, T, Jordan, K, Giesel, F. L, Villringer, A, & Der Radiologe, p. 552-557.

[19] Stöcklin, G. Nachr. Chem. Tech, Lab. 34; (1986). , 1057-1064.

[20] Wienhard, K, Wagner, R, & Heiss, W. D. In PET- Grundlagen und Anwendung der Positronen Emissions Tomographie, Springer Verlag Heidelberg; (1989).

[21] Coenen, H. H, & Der Nuklearmediziner, p. 203-214.

[22] Coenen, H. H. In Clinical Molecular Anatomic Imaging: PET, PET/CT and SPECT/CT, Ch.16, PET-radiopharmaceuticals: fluorinated compounds. Lippincott Williams & Wilkins; (2003).

[23] Defrise, M, & Gullberg, G. T. Image reconstruction. Phys Med Biol. 51; (2006)., 139-154.

[24] Lalush, D. A, & Wernick, M. N. Iterative image reconstruction, in: Wernick MN, Aarsvold JN, eds., Emission Tomography: The Fundamentals of SPECT and PET. 1st ed. San Diego, CA. Elsevier; (2004)., 443-472.

[25] Qi, J, & Leahy, R. M. Iterative reconstruction techniques in emission computed tomography. Phys Med Biol. 51; (2006)., 541-578.

[26] Reader, A. J, & Zaidi, H. Advances in PET Image Reconstruction. PET Clin 2; (2007)., 173-190.

[27] Stühler, E, & Merhof, D. Principal Component Analysis Applied to SPECT and PET Data of Dementia Patients- A Review, Principal Component Analysis- Multidisciplinary Applications, Dr. Parinya Sanguansat (Ed.), 978-9-53510-129-1InTech; (2012).

[28] Merhof, D. J Cereb Blood Flow Metab, January 31(1); (2011)., 371-383.

[29] Braem, A. Chamizo Llatas M., Chesi E. et al. Feasibility of a novel design of high-resolution parallax-free Compton enhanced PET scanner dedicated to brain research. Phys Med Biol 49; (2004)., 2547-2562.

[30] Zaidi, H, & Hasegawa, B. H. Overview of Nuclear Medical Imaging: Physics and Instrumentation in: Quantitative Analysis in Nuclear Medicine Imaging, Zaidi. H. (Ed), Springer, 100387238549

[31] Van Den, P. A, & Elsen, E. J. D. Pol, and M. A. Viergever. Medical image matching- a review with classification. IEEE Engineering in medicine and biology, (1993)., 12(1), 26-39.

[32] Antoine Maintz JB., Viergever Max.A. An Overview of Medical Image Registration Methods, European journal of nuclear medicine, 22(4); (1995)., 351-355.

[33] Hill Derek LG. et al. Medical image registration, Phys. Med. Biol. 46; (2001)., 1-45.

[34] Jannin, P, et al. Validation in Medical Image Processing, Medical Imaging 25(11); (2006)., 1405-1409.

[35] Dewinter, F, Gemmel, F, Wiele, C, Poffijn, B, Uyttendaele, D, & Dierckx, R. Fluorine fluorodeoxyglucose positron emission tomography for the diagnosis of infection in the postoperative spine. Spine. 28(12); (2003)., 1314-1319.

[36] Clinical Applications of SPECT/CT: New Hybrid Nuclear Medicine Imaging System IAEA-TECDOC-1597 ; August (2008).

Chapter 2

Assessment of Safety Standards of Magnetic Resonance Imaging at the Korle Bu Teaching Hospital (KBTH) in Accra, Ghana

Samuel Opoku, William Antwi and
Stephanie Ruby Sarblah

Additional information is available at the end of the chapter

1. Introduction

The various equipment and chemicals used in the radiology departments can be a source of hazards and hence result in an adverse effect to affected individuals (Johnston and Killion, 2005). Interdisciplinary approach to monitor the activities at radiology departments to ensure compliance in safety standards may help avoid or reduce hazards in the working environment (Byrnset al., 2000). Magnetic Resonance Imaging (MRI) unit in a radiology department is one particular place where safety precautions should be directed due to the ferromagnetic nature of the equipment and the strong magnetic field used in its operations (Joseph, 2006).

MRI is a painless, non-invasive and one of the most advanced imaging modalities currently available in radiology departments (Kusumasuganda, 2010). Research and awareness of safety issues concerning MRI has received much attention (Ordridgeet al., 2000). According to Westbrook et al (2009), recent occurrences in the operation of MRI have led to questions being raised on the safety of the modality. Phin (2001) has suggested that adequate policies and procedures should be developed and adhered to in order to ensure safe, efficient and operating conditions of MRI. Several potential problems and hazards are associated with the performance of patient monitoring and support in the MRI environment (Kanal and Shellock, 1992). According to Henner and Servomaa (2010), the main factors that affect safety practice in the MRI unit is management style and attitude of staff. Various reports found in the literature have indicated that MRI accidents are mostly caused by human errors rather than scanner malfunction. These have led to several calls for regulations and policies to guide the operations of MRI (New York Times, 2010; Healthcare Purchasing News, 2010). This has become necessary because the risk in the MRI environment does not only affect the patient, but also affects the health professionals and those who find themselves in the magnetic field (Kanal *et al.*, 2007). There is therefore the need for maximum safety to be ensured in the MRI unit. Moreover Chaljub (2003) and Joseph (2006) have both emphasized the need to keep training health personnel on safety issues relating to MRI. In particular, Joseph reiterated that the MRI's magnet which is over 100,000 times the earth's natural magnetic pull is always on mode; hence those who approach it should have training due to the special safety risk it poses. In addition to the risks to people, it is also important to put in precautionary measures to protect the equipment from damage and breakdowns. The need to assess the staff of the radiology department and hospital's management on their attitude and adherence towards maintaining safety at the MRI can therefore not be overemphased.

In recent times, Magnetic Resonance Imaging (MRI) unit of the Korle - Bu Teaching Hospital (KBTH) in Accra, Ghana has witnessed various degrees of accidents. In particular, there was a fire outbreak in 2007 which brought the operation of the MRI facility down for a whole year. Again in 2010, a wheelchair was pulled into the gantry of the MRI scanner by the missile effect when a patient was lifted off the wheelchair onto the MRI table as shown in appendix I on page 28. This incident resulted in three weeks down time of the facility. A second incident in the same year occurred where a Radiographer Intern at the MRI unit wrongly switched off the safety button, resulting in three weeks shut-down of the entire unit. These incidences have been documented in the Incidence Reporting Book at the MRI Unit and are reproduced here with the permission from the Radiology Department of the hospital. These incidences at MRI Unit at the Korle Bu Teaching Hospital are very worrying, suggesting that the safety aspects might have been compromised. Thus it is imperative that the existence of policy guidelines and manuals regarding the operational safety of the MRI in the hospital and their compliance and adherence by staff needed to be evaluated. Similar incidences occurring in other hospitals around the world are documented in the literature and some of which are reproduced in Appendix II on page 29.

2. Materials and methodology

This study was undertaken at the MRI Unit at Korle Bu Teaching Hospital in Accra, Ghana. The specific objectives of study were to identify safety policies regarding the operations of the MRI unit and whether they conform to international standards. Additionally, it sought to ascertain adherence and compliance of the policy guidelines and to evaluate the design features of the MRI suite for its safety compatibility as well as to determine the safety training needs of radiographers who operate the MRI.

The study focused on the safe use of MRI as an imaging modality and involved radiographers of the Radiology Department. A member of the Hospital Management Team also participated in the study since the management members are responsible for the safety policies formulation and ensuring their implementation at the MRI unit.

3. Literature review

In order to have a broad perspective of MRI safety issues, an extensive literature review were done which centred on the concept of safety screening; principle and framework of safety in MRI; operational principles, safety policies and guidelines of MRI.

3.1. The concept of safety screening

It has been suggested that in dealing with safety issues the emphasis should be placed on prevention of accidents (Harding, 2010). This means measures need to be implemented to prevent accidents from occurring. Harding argued that even though total prevention of accidents is not achievable, every effort should be made to reduce their occurrences to the barest minimum. The concept of safety has a wider significance as safety is seen as a systemic approach with thresholds that define the standard of safety (Elagin, 1996). In order to ensure an accident free, Elagin has suggested that an ordered procedure, which shows the level of safety in a particular environment should be followed. In recent times, concerns have been raised about the safety of the MRI facility due to the increasing number of MRI incidents by an alarming 185% over the last few years (Gould, 2008). Gould further suggested that there is need for a comprehensive safety programme for any health institution with a zero tolerance for MRI errors. Several studies have shown that compromising patient safety have resulted in fatal consequences (Launders, 2005; Emergency Care Research Institute (ECRI), 2004). In 2005, Launders conducted an independent analysis of the Food and Drug Administration (FDA)'s Manufacturer and User Facility Device Experience Database (MAUDE) and gave a report on a database over a 10-year time span. This revealed 389 reports of MRI-related events, including nine deaths with three events related to pacemaker failure, two due to insulin pump failure and the remaining four related to implant disturbance, a projectile, and asphyxiation from a cryogenic mishap during installation of an MR imaging system. Various claims have been made in several publications which indicate that MRI accidents are largely due to failure to follow safety guidelines, use of inappropriate or outdated information related to the safety aspects of biomedical implants and devices and human errors (Shellock and Crues, 2004; New York Times, 2010; Healthcare Purchasing News, 2010). A panel under the auspices of the American College of Radiology (ACR) was constituted to address these critical issues. Kanal et al (2004) who were part of this panel pointed out that there was a continuous change in the use of the MRI as a technology with a drastic increase in the number of examinations done. They maintained that though there were safety guidelines, the increased number of MR practitioners and the increased use of the technology for critically ill patients, contributed to the increasing incidence of mishaps occurring in MRI surroundings. According to McRobbie et al (2007), the overall objective of a safety procedure is to provide an appropriate standard of protection of patients and staff in the MRI unit, without unduly limiting the beneficial practices and also prevent the occurrences of tragic events in the MRI suite. MRI suites in clinical and hospital surrounding should establish safety protocols with an MRI safety officer designated to ensure that policies are implemented and adhered to (Kanal, 2004).

3.2. MRI suite design and zones

An MRI suite should be designed to restrict access and limit exposure to static magnetic fields. Various publications have provided different designs to the MRI suite to ensure maximum safety and they all showed that an MRI suite should be built to restrict access by zones (Gould 2008; Kanal *et al.*, 2007; Junk and Gilk, 2005; Shellock and Crues, 2004). The zones suggested by the various articles are as follows;

Zone I: - Opens to the general public and presents the least exposure to the patients, staff and visitors. Usually it is the reception and waiting room for the MRI suite purposed to channel patients and medical staff to the pre-screening area (zone II) and limit entry to the MRI suite.

Zone II: - This is the first interaction site for patients, visitors and staff in the MRI suite. The purpose of this zone is to restrict further public access to the suite and provide direct supervision of patients and visitors by the MRI staff. Pre-screening of all patients, staff and visitors also takes place here. If ambulatory, the patient is screened through a ferrous metal detector installed into the zone II. Non ambulatory patients in walkers, wheelchairs or patient support need the transport equipment to be verified as MRI- safe or exchanged for MRI- safe equipment. The zone II generally has a metal detector and a 1000 gauss magnet to help screen medical equipment for ambulatory patients. MRI staff including the MRI technologist is directly responsible for enforcing strict adherence to the MRI safety protocols for the MRI suite and patient safety.

Zone III: - This is the entry zone to the MRI machine room which is zone IV. Entrance to this zone is restricted physically and by protocol. Being the last barrier against an incident or injury due to an interaction of a static or active magnetic field and any unscreened personnel, patient or equipment, only MRI technologist, certified staff and pre-screening attending physician accompany the patient into the MRI machine room. The portal or entrance to the MRI machine room must be monitored by a second ferromagnetic- sensitive detector and door must be locked. Sounding of detector will require verification of either an MRI- safe or compatible event or the discovery of an MRI–unsafe condition in the patient, transporting or medical equipment or the attending medical staff. The standard access method is a card access system which should allow access to only certified MRI staff between zone III and zone IV. All medical staff must be pre-screened prior to entry into zone III to make sure no unscreened individuals will be allowed access to zone IV. Ideally, the personnel in zone III must be uniformed in MRI compatible scrubs which will avoid the use of identification badges in the suite, MRI-safe shoes and undergarments. Personnel must avoid all jewellery, watches metallic writing instruments, and wire-framed glass which may raise a false alarm from the detector (Shellock and Crues, 2004).

Finally in zone IV, the MRI room should have a clear demarcation of the five gauss line taped or painted on the MRI suite floor to indicate the area beyond which requires MRI- safe or MRI – conditional equipment or instrumentation. This should be in line with the distance and tesla rating of the MRI. Zone IV should be clearly marked with a red light and lighted sign stating, "The Magnet is on." In situations where an alarm goes off for a code red, there is the need to use MRI- safe equipments to address the situation with restrictions of public first responders from zone IV of the MRI environment until safe conditions are established or responders are verified to be safe (Junk and Gilk, 2005). Access to zone IV should be enabled by a programmed key and the key kept in a restricted access box in the MRI control room.

Architectural and design engineering for a MRI suite have been established in standards published by the Joint Commission on the Accreditation of Health Organisations (JCAHO), the American College of Radiology guidelines, the International Building Code (IBC) and Occupational Safety and Health Administration (OSHA).

3.3. Pre- screening and screening forms

Shellock and Crues (2004) emphasized that the establishment of thorough and effective screening procedures for patients and other individuals is one of the most critical components of a programme to guard the safety of all those preparing to undergo MR procedures or to enter the MR environment. All preliminary patient history, MRI safety screening and documentation must be completed and signed by the patient, guardian or clinician before procedures are undertaken on patients (Shellock and Crues, 2004; Ferris *et al.*, 2007). Various screening forms are used for different categories of people who come to the MRI suite. In general, screening forms are developed with patients in mind (Sawyer-Glover and Shellock, 2002).

3.4. Colour and symbol coding

Various means have been adopted to help with ensuring safety in an MRI unit. This is usually meant to provide on the spot recognition of MR- safe equipment and surroundings, likewise unsafe and MRI-conditional equipments and locations. At the University of California San Francisco (2011), yellow is used to signify caution and is painted around the entrance of the door. Gas tanks that have been painted green signifies ferrous cylinder and hence make it easy for identification as MRI unsafe equipment. For safe MRI tanks, the cylinders are coloured silver. Symbols have also been used as a new classification system for implants and ancillary clinical devices.

An MRI safe symbol signifies that the device or implant is completely non-magnetic, non-electrically conductive, and non-RF reactive, eliminating all of the primary potential threats during an MRI procedure. An MRI Conditional sign is used to identify a device or implant that may contain magnetic, electrically conductive or RF-reactive components that is safe for operations in proximity to the MRI, provided the conditions for safe operation are defined and observed (such as 'tested safe to 1.5 teslas' or 'safe in magnetic fields below 500 gauss in strength). Finally, an MRI unsafe symbol is reserved for objects that are significantly ferromagnetic and pose a clear and direct threat to persons and equipment within the magnet room. An appropriate coding system is thus necessary to be adopted by every MRI unit to facilitate easy identification of safe items.

3.5. Operational principles of MRI

As opposed to conventional x-rays and computed tomography (CT) scans, there is no ionizing radiation used in MRI. However, MRI uses an extremely powerful static magnetic field,
rapidly changing gradient magnetic fields and radiofrequency electromagnetic impulses to
obtain detailed anatomic or functional images of any part of the body (Faulker, 2002; Berger,
2002). Currently, there is no evidence of a short or long term adverse effect due to exposure
to field strengths of MRI and durations that is clinically used (Schenck, 2000).

Despite the relative safety of MRI, there are potential hazards associated with its operations.
Some of these are related to the physical properties of the MRI equipment and also to the
challenges of maintaining physiologic stability of the individual undergoing the examination. In a reported incident in 2001,a small boy undergoing an MRI following surgery to remove a benign tumour was struck and killed by an oxygen tank inadvertently taken into the
MRI suite (Emergency Care Research Institute, 2001). In most situations the MR systems
cause the disaster due to it interactions with other properties around it.

3.6. Magnetic fields and the missile effect

The static magnetic field generated by a powerful magnet is tens of thousands times
stronger than the earth's magnetic field which can attract objects containing ferrous materials, transforming them into dangerous airborne projectiles (Dempsey et al., 2002). There
are two features of the magnetic field that are the source of most MRI incidents; the projectile or missile effect which is the ability of the magnet to attract ferromagnetic objects
and draw them rapidly into the bore with considerable force (Centre for Devices and Radiological Health, 1997). Ferromagnetic objects include metallic objects containing iron
such as scissors, laryngoscopes, nail clippers, pocket knives and steel buckets. Larger
items like wheelchairs, gurneys, intravenous poles have also become MR-system- induced
missiles (Centre for Devices and Radiological Health, 1997). The other source of most MRI
incidents is the translational attraction which occurs when one point of an object in a
magnetic field is attracted to a great extent than the object's furthest point from the at-
tracting source (Gould, 2008).

3.7. Magnetic field interactions

The static magnetic field of an MR system is always on. No sound, sight, smells alerts personnel to the presence or the extent of the invisible field surrounding the magnet in all directions. The magnetic pull is strongest at the centre of the MR system and weakens with
increased distance from the magnet, creating a spatial magnetic field gradient (Price, 1999).
The distribution of the magnetic field outside the main magnet called fringe field is impossible to see, but it is critical to safety in the MR environment because it can determine whether
a ferromagnetic object could become a projectile. MR systems with large fringe field generally create the greatest hazards (Price, 1999). If the fringe strength decreases more gradually
with distance from the magnet, the object's attraction to the magnet progressively strengthens as it becomes closer to the magnet. Personnel within the MR room may notice an in-

creasingly stronger pull on objects they are wearing or carrying as they walk closer to the MR system, permitting them to retreat from the MR system before an accident occurs (Kanal et al., 2002).

3.8. Hazards in the MRI suite

Various forms of hazards occur in the MRI suite which can be categorized into translational force- missile effect, torque forces, induced magnetic fields, thermal heating and quenching (Colletti, 2004). In the translational force, the effect is manifested on the ferromagnetic materials and the static field generated by the MR system usually in the form of the missile effect involving non-compatible objects and miscellaneous patient and visitor objects.A hair or paper clip within the 5-10 gauss line range could reach a velocity of 40 mph (about 70 kph) and will be attracted to the centre of the lines of force of equal (Lahr and Rowan, 2004).

Just like the translational forces, the torque force is also associated with ferromagnetic materials and the static field generated by the MR machine. Ferromagnetic objects that are attracted by the magnetic field react by aligning parallel to the magnetic lines of the force being created by the MRI machine. The centre of the MRI- generated fields has the highest torque force, creating a serious exposure for all contraindicated items and MRI- conditional items in the MRI suite, depending on the tesla rating of the MRI (Gould, 2008). When any metallic object is introduced into a high flux field, current will be induced if that object is perpendicular and moving to the lines of the force. The new current will create a secondary magnet field that will oppose the original field. This can cause patient discomfort and anxiety due to the reactive forces on the MRI safe medical implants and a life threatening condition may be created under the five- gauss line (Kangarlau and Robitaille, 2000).

Magnetic Field type	Hazard	Potential Adverse Effects
Static magnetic field	Translational force: power. Attraction of ferromagnetic objects to intense magnetic field. Rotational force/ torque: rotation of object to align with the magnetic field	Missile effect: acceleration of objects into the bore of the magnet. Tearing of tissues, pain, and dislodgement of some implants.
Radiofrequency electromagnetic fields	Heating due to absorbed RF energy Electromagnetic interference	Overheating, burns (thermal, electrical) Device malfunction; imaging artefact
Gradient magnetic field	Induced currents in conductive tissues Induced current in electrical devices	Nerve and muscle stimulation Device malfunction/failure

Adapted with permission from Centre for Device and Radiological Health of USA

Table 1. Hazardous Magnetic Field Interactions

The most common source of thermal exposure tends to be looped or un-looped medical equipment leads, MRI accessories and sensors. The most serious exposure is located in the bore of the MRI machine and in the axis points, as they possess the highest potential torque forces. Extremity coils could increase the risk but this can be avoided by the use of MRI safe polymeric foam padding (Gilk, 2006). MRI machines are cooled by a super cooling fluid (liquid helium). The release of the super cooling fluid into the atmosphere is called quenching. Most clinical machines have about 700 to 1000 litre volume of this cryogenic. In the event that there is venting, it may cause the oxygen in the MRI room to condense around the vent pipe and accumulate in the MRI machine causing a red fire hazard. Another risk is a quench vent pipe breech which could flood the room with cryogenic fluids creating an asphyxiation hazard for the patient and the staff (Clark, 2007).

3.9. Radiofrequency electromagnetic fields effects

The MRI system has electromagnetic coils in a transmitter within it that delivers the radiofrequency (RF) pulses during imaging. When tissues absorb the RF energy, tissue heating can occur, mostly in patients with poor thermoregulatory control (Dempsey *et al.*, 2002). The rate at which RF energy is deposited in tissue is known as the specific absorption rate (SAR), measured in units of watts per kilogram (w/kg) (Centre for Devices and Radiological Health, 1997). The maximum allowed SAR is 3W/kg which is averaged over ten minutes for head imaging and 4W/kg for whole body imaging, averaged over fifteen minutes (Centre for Devices and Radiological Health, 2003).

Radiofrequency fields can cause skin burns if monitor cables or wires are permitted to form conductive loops with themselves or with other body parts (Kanal *et al.*, 2002). Temporary metallic intra cardiac pacing wires will behave like antennae and conduct electromagnetic waves, also resulting in thermal tissue injury (Dempsey *et al.*, 2002). Radiofrequency signals emitted during the MR examination can affect non- MR-compatible programmed infusion pumps, resulting in erratic performance. Affected pumps could deliver higher or lower than desired volumes of pressor agent, analgesics, sedative or dextrose and electrolytic solutions, all of which cause serious physiological consequences particularly, infants (Cornette*et al.*, 2002).

3.10. Gradient magnetic field effects

When an infant is subjected to sudden, rapidly changing gradient magnetic fields during imaging, the magnetic field can induce circulating currents in conductive tissues of the body (Schaefer *et al.*, 2000). These currents have been found to be large enough to produce changes in nerves and muscles function theoretically.Where safety standard limits are practiced, it limits the maximum rate of change of magnetic field strength that can be used thus reducing the likelihood of its observation during a clinical MRI (Center for Devices and Radiological Health, 2003).

3.11. Safety policies and guidelines of MRI

The American College of Radiology (ACR) Guidance Document for Safe MRI Practices-2007 recommends that all MRI sites should maintain MR safety policies (Kanal *et al.*, 2007). These policies, it claims should be reviewed concurrently with the introduction of any significant changes in the safety parameters of the MR environment and updated as needed. It also stated that Site Administration is responsible to ensure that the policies and procedures are implemented and adhered to by all site personnel. Any adverse events, MR safety incidents or near incidents are to be reported and used in continuous quality improvement efforts. To augment the recommendations made by the ACR, the 2008 Joint Commission Sentinel Alert issued by the Medical College of Wisconsin's (2009) accreditation organisation suggested that actions consistent with the ACR recommendations should be used to prevent accidents and injuries in the MRI suite. In other works, the Device Bulletin (2007) produced a document to serve as guidelines covering important aspects of MRI equipment in clinical use with specific reference to safety. They were intended to bring to the attention of those involved with the clinical use of such equipment, important matters requiring careful consideration before purchase and after installation of the equipment. It was also to be used as an orientation for those who are not familiar with the type of equipment and act as a reminder for those who are familiar with the equipment (Buxton and Lui, 2007). It was further intended to act as a reminder of the legislation and published guidance relating to MRI, draw the attention of the users to the guidance published by the National Radiological Protection Board (NRPB), its successor the Health Protection Agency (HPA), the International Electrochemical Commission (IEC) and the International Commission on Non –Ionizing Radiation Protection (ICNIRP)

4. Materials and method

The study employed both qualitative and quantitative design using a structured interview and descriptive survey. A structured interview involves guiding the interview in a particular pattern such that the information received falls in line with the objective of the study without it being altered by the interviewer (Brink and Wood, 1994; Pontin, 2000). A descriptive survey provides a better means of investigating and assessing the attitude and practices of people when they are involved in a particular situation (Carter, 2000; Gray, 2004).

The study was carried out MRI Suite of the Radiology Department of the Korle Bu Teaching Hospital. (KBTH), Accra. Ghana. KBTH is the leading referral hospital in Ghana, with the radiology department being one of the busiest departments in the hospital. Currently, the hospital has a bed capacity of about 2000, with an average 1,500 outpatient attendances daily, an admission rate of 250 per day and 65% of the daily attendance visiting the radiology department (www.korlebuhospital.org).

The Radiology Department of the hospital has a staff population of forty-six. These include thirty-one radiographers, nine radiology residents and six consultant radiologists. Of the thirty one radiographers, twelve are degree holders, fifteen are diploma holders and the rest are certificate holders.

The entire population of radiographers was used for the study. This gave a population size that was easy to handle and ensure an effective statistical analysis to be done (Burns and Grove, 2001). Using a small data set makes it possible to overcome the inconveniences created by lack of time, ensures homogeneity, improves the accuracy and quality of the data (Atkinson, 2000; Aderet al., 2008). In Korle Bu Teaching Hospital, there are no specialized or permanent MRI radiographers and all of them rotate periodically to the MRI unit; hence the reason for using the entire population of radiographers for the study.

Polgar and Thomas (2000) emphasized that in any scientific research the primary consideration is the protection of the rights and welfare of participants. Thus, ethical approval was sought from the Ethical Review Committee of the School of Allied Health Sciences, College of Health Sciences, University of Ghana. Permission was also sought from the Dr. Frank G. Shellock (2002) to reproduce content in his work and from the Institute for Magnetic Resonance Safety, Education and Research as well as the Radiology Department of the Korle Bu Teaching Hospital for the use of facility for the study. Informed consent was sought from participants in the form of written consent forms after the objectives of the study had been explained to them. They were assured of their anonymity, confidentiality of identity and information provided.

A self- administered open and closed ended survey questionnaire was used to obtain data from the participants. Before the main study, a pilot study involving three radiographers was conducted to assess the validity and reliability of the questionnaires. A modified checklist designed by Gillies (2002) was attached to the pilot phase for respondents to make suggestions that helped to modify the questionnaire as required. Ambiguity was thus removed; clarity of the format and design adopted was also ensured (Bailey, 1997).

A structured interview also was conducted with a member of the hospital management. This was to obtain additional data, validate and verify results obtained from the survey (Polgar and Thomas, 2000). Policy formulation and supervision of implementation is the responsibility of management of the hospital and the department (Beddoe et al., 2004). Thus interviewing members of the management was considered the best way to obtain detailed and comprehensive information about safety management.

Questions on the framework of operational safety of the MRI unit, training programmes and practical safety problems faced by the MRI unit were among other things asked during the interview. The interview was electronically recorded, transcribed and data grouped into themes and analyzed.

The data obtained from the questionnaires was rearranged in an ordered manner to enhance its processing by the Statistical Package for Social Sciences (SPSS) version 16.0. Nominal and ordinal levels of measurement were used because the study design was a descriptive survey (Burns and Grove, 2001). Results were presented using descriptive statistics in the form of charts, frequency tables and percentages.

5. Results

This study investigated the availability of safety policies and guidelines and adherence to them by staff at the MRI suite of the Korle-Bu Teaching Hospital. It also investigated the design features of the MRI suite as to whether it meets the acceptable safety standards. A total of thirty-one closed ended questionnaires (31) were distributed to all the practicing radiographers twenty eight (28) questionnaires were completed and returned completed giving a response rate of 90.3% (n=28/31).

Gender of Respondents	Professional qualification held			Total
	Certificate	Diploma	Bachelor's degree	
Male	2	10	9	21
	7.1%	35.7%	32.1%	75.0%
Female	2	2	3	7
	7.1%	7.1%	10.7%	25.0%
Total	4	12	12	28
	14.3%	42.9%	42.9%	100.0%

Table 2. Demographic Data of the Respondents

The profile above shows that the ratio of male to female respondents was 3:1. Nearly half of the respondents (42.9%) were both diploma and degree holders respectively

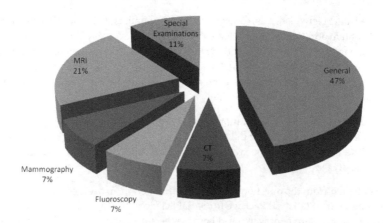

Figure 1. Area of work of Respondents. As shown, majority (47%) of Radiographers were engaged in general radiography.

Question Code	QUESTIONS	Responses to MRI Safety issues		Total
		Yes	No	
Q1	Is there a restricted access to everyone who comes to the MRI Suite	26 (92.9%)	2 (7.1 %)	28 (100.0%)
Q2	Do you undertake screening of patients who enter the MRI Suite?	21 (75.0%)	7 (25.0%)	28 (100.0%)
Q3	Do you undertake screening of staff who enters the MRI Suite?	10 (35.7 %)	18 (64.3 %)	28 (100.0%)
Q5	Are there lockers to store personal belongings that may be ferrous in nature or has a magnetic stripes in the MRI Unit	16 (57.1%)	12 (42.9%)	28 (100.0%)
Q6	Is the equipment used in the MRI environment checked by any authority, deemed MRI safe and labelled as such prior to implementation?	21 (75.0 %)	7 (25.0%)	28 (100.0%)
Q7	Does the equipment used in the MRI unit have colour codes to identify ferrous material and MRI safety material?	5 (17.9%)	23 (82.1)	28 (100.0%)
Q9	Do patients complete any MRI history and assessmentform that addresses possible contraindications prior to any MRI procedure?	26 (92.9%)	2 (7.1%)	28 (100.0%)
Q11	Are there proximity access doors and emergency exits to MRI suite?	15 (53.6%)	13 (46.4%)	28 (100.0%)
Q12	Do you face any problems in your bid to ensure the safety of patients and staff in the MRI unit?	14 (50.0%)	14 (50.0 %)	28 (100.0%)
Q14	Is there an assigned anaesthetist to the MRI unit to undertake procedures that need patients to be anesthetized?	7 (25.0%)	21 (75.0%)	28 (100.0%)

Table 3. Responses to MRI Safety Issues

In Table 3, majority of 92.9 % (n=26/28) of the respondents stressed the need to restrict access to the MRI suite. Additionally, 75.0 % (n= 21/28) of the respondents were of the view that patients should be screened before allowed to enter MRI suites. However, only ten out of twenty eight (n= 10/28= 35.7%) suggested screening for radiographers' (workers), before they enter the MRI suite.

Table 3 also shows that 57.1% (16/28) of the respondents mentioned the presence of lockers in the MRI suite to store personal belongings that may be ferrous in nature or has magnetic stripes. Furthermore, 75% (n=21/28) reported that equipment used in the MRI environment undergo regular quality check. However, 82.0% (23/28) of the respondents disclosed that the equipment in the MRI unit did not have colour codes to identify ferrous material and MRI safety material.

Majority of the respondents (92.9%) reported that prior to procedure or examination, patients are asked to complete questionnaires to determine any contraindications they may have. A significant majority (53.6%) reported the presence of emergency exits to the MRI suite. Finally 75.0% of the respondents were unaware of availability of anaesthesia services at MRI unit for patients who would require anaesthesia as part of the procedure.

Question Code	QUESTIONS	Responses to Safety MRI training and unit design features		Total
		Yes	No	
Q15	Have you had any training programme(s) on MRI safety issues?	6 (21.4%)	22 (78.6%)	28 (100.0%)
Q16	Have you attended any of such training programmes, if yes to question 18?	4 (14.3%)	24 (85.7%)	28 (100.0%)
Q17	Do you know the MRI zones?	8 (28.6%)	20 (71.4 %)	28 (100.0%)

Table 4. Responses to Safety MRI Training and Unit Design Features

Training gap was identified in the use of MRI equipment and safety as demonstrated in Table 4 71.4% of the respondents were not aware of the existence of any MRI zones in the suite.

Table 5 shows that in general, the majority of the responded were unaware of the basic knowledge about the operations of MRI.

5.1. Analysis of the open ended part of the survey questionnaire

With regard to procedures undertaken for patients and staff entering the MRI unit, the majority of respondents (67.86%) had very little knowledge about what were being done. The remaining 32.14% had a fair idea but could not provide detailed description of the exact steps that were undertaken. When it came to the colours used to identify ferrous and MRI safe material, there was a poor appreciation, evident by the fact all the 28 respondents did not know the existence of the colour identification of ferrous and MRI safe material.

QUESTIONS	Responses to MRI General Knowledge					
	Strongly disagree	Disagree	Undecided	Agree	Strongly agree	Total
Q21.The magnet is only on during the working day	14 (50.0%)	2 (7.1%)	3 (10.7%	4 (14.3%)	5 (17.9%)	28 (100.0%)
Q22.A strong magnetic field produces X-ray used for imaging	13 (46.4%)	2 (7.1 %)	1 (3.6%)	4 (14.3%)	8 (28.6%)	28 (100.0%)
Q23.A static magnetic field strength may be up to 100,000 times the magnetic field strength of the earth	1 (3.6%)	6 (21.4%)	13 (46.4%)	4 (14.3%)	4 (14.3%)	28 (100.0%)
Q24.5 Gauss line is the parameter around the MRI system where field strength is over 5 Gauss	0 (0.0%)	2 (7.1 %)	22 (78.6 %)	4 (14.3%)	0 (0.0 %)	28 (100.0 %)
Q25.At 5Gausspacemakers may be affected, ferrous items become potential flying projectiles and magnetic stripes are erased	1 (3.6 %)	1 (3.6 %)	19 (67.9%)	4 (14.3 %)	3 (10.7 %)	28 (100.0 %)
Q26.Below 5 Gaus is considered to be a safe level of magnetic field exposure to the public	0 (0.0%)	2 (7.1 %)	20 (71.4 %)	2 (7.1 %)	4 (14.3%)	28 (100.0%)
Q27.Magnetic field strength is measured in Tesla (T) or Gauss (G)	2 (7.1 %)	2 (7.1 %)	3 (10.7 %)	8 (28.6 %)	13 (46.4 %)	28 (100.0 %)

Table 5. Responses to general knowledge about MRI

There was, however, high knowledge level in relation to how patients with implants and ferrous materials around the MRI suite were managed evident by the that fact 64.29% respondents could describe the correct steps that should be taken for such patients.

All the respondents cited the following as challenges in ensuring safety of patients and staff at the MRI unit;

- A communication gap between patients and health professionals

- Unwillingness of co-workers to comply with protocol used at the unit

- Small waiting area which is shared by the CT-scanning unit and the MRI unit

- The lack of knowledge on continuity of procedure that has been performed for patients by referring clinicians.

On the issue of zoning in an MRI suite, 14.29% were aware of the different zones that are needed in a standard MRI unit as against 96.4% of the respondents which were not aware of the colour used to indicate the different zones. According to 21. 43% of the respondents, zoning was completely absent at the MRI unit of the Korle Bu Teaching Hospital. The general overview of the results suggests a huge knowledge gap on the safety issues of MRI by majority of the respondents.

5.2. The interview data

Thematic analysis was used to analyse the qualitative interview data. The predominant themes that emerged were the context for framework for operational safety at MRI unit, availability of departmental policy manual and training programmes for MRI. The areas identified included a maintenance programme for the MRI unit, structures in place for accidents and breakdown of the MRI unit and practical problems faced in ensuring safety at the MRI unit.

The context for framework for operational safety at MRI unit in this study represents a combination of organisational and operational methods that from the radiographers perspective, significantly affect the achievement of operational safety at MRI unit. It was noted that MRI was a relatively new modality in the country with the suite at the Korle Bu Teaching Hospital which was in 2006 being the first in the country. It was further observed that there was no documented formal framework by either the hospital or the department and that preparations were underway to produce one in accordance with best international standards. This observation was consistent with the findings from the survey questionnaire indicating the absence of a policy manual at the radiology department.

On the issue of continuous education and training of radiographers on MRI, it was noted that this was non-existent. As stated earlier, there was no MRI specialized radiographer in the department and any qualified radiographers who has basic knowledge in MRI could be assigned to the unit. The need to provide a platform to training and educate the practicing radiographers on MRI was identified.

On maintenance programme for the MRI unit, it was mentioned that the supplier of the equipment has a maintenance contract with the hospital to undertake routine maintenance of the MRI.

On measures that were being taken to prevent further accidents, it was indicated that the staff were required to report any incident or missed-incident to the appropriate authority. The absence of zoning in the suite and the adjoining CT scan suite made it difficult to undertake any effective screening because both MRI and CT scan patients have to enter through the same entrance. The absence of access codes for entrance into the MRI unit was also identified as a challenge to restricting access and this was attributed to defect in MRI – suite.

Other challenges identified include the attitude of some hospital personnel who were not willing to comply with safety and security measures in place. The need to undertake some structural adjustment to the unit was being considered to detach the CT suit from the MRI. Ensuring total commitment from both the management and staff of the hospital to safety and security issues was identified as one of the main means to prevent accidents at the MRI unit.

6. Discussion

This research sought to investigate the availability of safety policies and guidelines and adherence to them by staff at the MRI suite of the Korle-Bu Teaching Hospital. It also investigated the design features of the MRI suite to ascertain whether it meets the acceptable international safety standards as these inevitably, affect patient care during MRI procedures. In this chapter, the findings are discussed and key issues which require immediate attention are identified.

6.1. Response rate

Out of the 31 questionnaires administered, 28 were returned providing an appreciable response rate of 90.3% (n=28/31). The high response rate received could be due to the small population of radiographers in the department and their easy accessibility. It could also be due to the time and period that the data was collected; just after close of work. A response rate above 50% is an important part of a survey because it enables findings to be generalized (Burns and Grove, 2003). The survey undertaken can thus be generalized to the population that was studied.

6.2. Demographic profile of respondent: gender, professional qualification and working area

The demographic profile in table 2 on page 13 shows that the ratio of male to female respondents was 3:1 (75% - 25%). This observation may be associated to the general perception individuals have of radiation. As espoused by Maiorova et al (2008) most females prefer to work in other professions than to be in the radiography profession which is consistent with the Ghanaian situation where high numbers of females are found in other professions, particularly, nursing. As a result of the misconception people have about radiation in Ghana, some nurses even refuse to stay in the duty room at the radiography department to assist patients that they have accompanied. However, in other parts of the world, especially Australia, the radiography profession is dominated by the female population (Merchant et al., 2011).

Table 2 on page 13 also shows that equal numbers of the respondents were either diploma or degree holders (42.9% each) with the certificate holders being the least (14.3%). This is due to the fact that the certificate programme had been phased out long time ago and recently the diploma programme has also been stopped. The only radiography educational programme currently being offered in Ghana is the bachelor's programme. It was however observed that there was no respondent with a postgraduate degree hence the highest educational qualification in the study setting was first degree holders. As a result the absence of post-graduate education in the country, only few radiographers have managed to acquire post graduation abroad and they are mostly in the academia.

Figure 1 on page 13 shows that a good number of the respondents (47%) were into general radiography. The increased requests for general radiography examinations and the increased number of duty rooms may be responsible for this trend. Facilities for specialised imaging modalities are very limited and as has been stated earlier, there is only one MRI, a CT-scan and one mammogram in the department, hence the majority of the respondents in general radiography.

6.3. MRI safety issues

In this study, majority of the respondent were of the view that access to the MRI suite should be restricted to everybody who enters the unit. Whilst majority of the respondents claimed that patients were screened before entry, the same could not be said about the staff members as only ten respondents reported to be screening staff members. However, this assertion could not be entirely true because close observation during the study revealed that not a single staff was made to undergo mandatory screening apart from taking out their metallic possessions on their own volition. This also goes to confirm the assertion that the personnel were unwilling to comply with safety and security protocols at the unit. This assertion was corroborated during the interview about the absence of coded access keys for staff in accordance with international best practices found in the literature (Kanal et al., 2007; Junk and Gilk, 2005; Shellock and Crues, 2004).It was also observed that patients were only made to change into gowns placed in the changing room of the MRI unit and all metallic opacities removed from them before entering the scanning room. A metal detector screening coupled with visual observation was the only form of screening that was done at the unit aside patients filling out an MRI screening form. There were no in - built detectors in the building to give off any alarm as an indication of the presence of a metallic substance (Gould, 2008). The study also showed that there were no lockers for both staff and patients to keep their valuables that may be ferrous in nature even though 57.1% of the respondents claimed that there were such facilities. The only available option for the staff was to keep their items in the rest room or bring them to the control panel area for safekeeping; which is not completely safety - assured. With regards to patients, their valuables were either kept in the changing rooms or brought to the control panel area. A positive observation made was the availability and use of an MR-compatible wheelchairs and trolleys. This development may be described as the reactive response by the management to the wheelchair incident that occurred at the unit as captured in the problem statement.The staff were also more vigilant and non ambulatory patients were thus transferred onto MRI safe wheelchairs and trolley before being sent to the scanning room.

It was reported by 57.1% of the respondents that the MRI equipment was regularly checked by the hospital authorities. However, it was established that these checks were not regular according to the standardized quarterly quality checks and maintenance scheme. The irregularity of the quality assurance checks could be a contributing factor to the frequent breakdowns of the equipment a view held by the respondents as contained in Table 3 on page 14

It was observed that patients were asked to complete MRI history and assessment forms to determine if they have conditions that were contraindicated to MRI procedure. However, non- patients including referring clinicians entering the unit did not complete this form. With the exception of the screening forms, no other safety and security documentation for both patients and staff were available in contravention of standardised policies and guidelines (Ferris *et al.*, 2007).

It was found that occasionally anaesthetists were assigned to the MRI unit for required procedures. This could explain why majority of the respondents were unaware of the presence of anaesthetic services at the radiography department.

6.4. MRI training and unit design features

The study revealed (as shown in table 4 on page 15) that there was a huge training gap in the use of MRI equipment. This was evident from the low general knowledge in MRI exhibited by the respondents, which was collaborated during the interview. The knowledge defect was also demonstrated by the fact that most of the respondent did not provide accurate responses to the questionnaires, a situation which may be attributable to the lack of policies and guidelines.

The study also found out that the design of the MRI suite did not conform to the basic design feature of a well laid out MRI unit as described by various organisations including the Joint Commission on the Accreditation of Health Organisations (JCAHO); International Building Code (IBC); Occupational Safety and Health Administration (OSHA)}. The defect in the design of the unit may be as a result of its mergence with the Computed Tomography (CT-scan) unit and other imaging modality units. The old CT- scan unit was collapsed and expanded to make room for the MRI unit and other imaging units thus preventing the ideal design of an MRI unit to be built out.

6.5. General knowledge about MRI

The responses on the general knowledge on MRI confirmed the training defect. It is possible that the few radiographers who had some knowledge about MRI acquired it through personal effort and on the job observations. Thus the absence of a framework for operational safety of the MRI could be a major issue that militates against the effective practice of safety at the MRI unit in the radiology department of the hospital.

6.6. Summary

Safety of patients and staff around the MRI unit is a critical issue in the practice of diagnostic radiology due to the high magnetic fields and radiofrequencies associated with the operations of the MRI scanner. Magnetic field associated with the MRI scanner is 10,000 times higher than the earth's magnetic field; therefore a detection of the smallest amount of ferrous in any material is essential. It is therefore essential that radiographers take practical steps to identify any unknown material in or on any patient or staff that may be ferrous in nature or magnetic-sensitive.

The creation of an attitude of safety screening, however, requires a firm commitment of both senior management and staff of the hospital, which must be communicated through policies and local rules.

7. Conclusions

7.1. Based on the findings of the study, the following conclusions are drawn;

• Poor documentation of safety issues at the department was noted.

• Safety screening was practiced to some extent but there were no written local rules or policies that actually specify what a radiographer should do routinely. There was therefore no standard of practice in the department.

• The safety screening undertaken in the MRI unit was done primarily on patients, overlooking the risks posed by other individuals and co - workers who come to the MRI unit

• There was lack of an effective and efficient policy and guidelines in the hospital in general and the radiography department in particular.

• The inappropriate design feature of the MRI suite was also seen to be a hindrance to effective safety screening practices.

• This research is the first of its kind to be conducted at the MRI unit of KBTH. It is our considered view that further work needs to be carried out to validate the assumption that the frequent accidents and breakdowns at the MRI unit is as a result of the lack of safety policies and operating guidelines at the unit. It would also me necessary to extend such a study to the other MRI units in the country to determine their safety and security levels,

7.2. Limitations of the study

This study was conducted exclusively in the Korle Bu Teaching Hospital with a study population of thirty one radiographers. Although KBTH is the leading referral hospital in Ghana, making generalizations about radiographers nationwide has to be done with caution since the sample may not be truly representative of the entire population. There are a very limited number of MRI scanners in the country with the one at Korle – Bu Teaching Hospital being the first. Hence this might not reflect the safety practice that take place at the other units since the other few may have the different designs which may meet international standards

7.3. Recommendations

• As a matter of urgency, professional bodies and Korle Bu Teaching Hospital should collaborate to produce a framework for the operational safety of MRI unit for the radiology department. In this framework, the department should come out with policy manuals and guidelines which would include specific safety issues which relate to the Ghanaian setting, training programmes to enhance the knowledge base of the radiographers. This should be reviewed regularly to meet the rapid advancement to the MRI technology.

• To equip the radiographers with practical experience in the use of MRI, the periodic rotation should be effectively implemented or more radiographers should be encouraged to upgrade themselves in the operations of MRI.

• The curriculum of the Diagnostic Radiography programme of the University should be thoroughly reviewed to cover the operational safety issues of the MRI. The practical examinations conducted during the final year should include all aspect of the medical imaging modalities and not only the conventional radiography. This is will equip the students with adequate practical experience of all the imaging modalities.

Appendix I

A photograph of the wheelchair that got trapped in the gantry of the MRI Scanner on the 12th of May, 2010 at the Korle Bu Teaching Hospital, Accra. Ghana

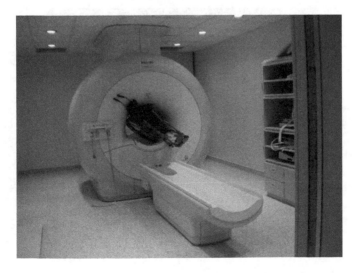

Figure 2.

Appendix II

MRI Incidents in different parts of the World

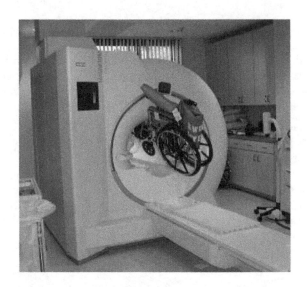

Figure 3.

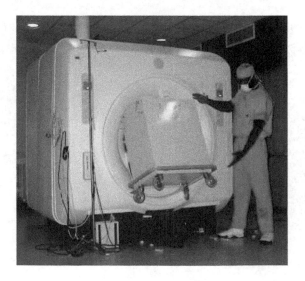

Figure 4.

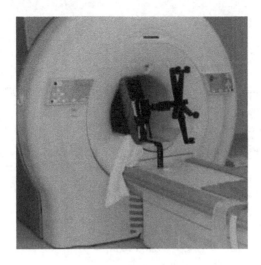

Figure 5.

Figure 6.

Author details

Samuel Opoku, William Antwi and Stephanie Ruby Sarblah

*Address all correspondence to: syopoku@chs.edu.gh

Department of Radiography, College of Health Sciences, SAHS, University of Ghana, Legon, Accra, Ghana

References

[1] Ader HJ, Mellenbergh GJ and Hand DJ. Advising on Research Methods: A consultant's companion, Huizen: Johannes Van Kessel; 2008.

[2] American College of Radiology. Glossary of MR terms (4th edn). Reston: VA; 1995

[3] Atkinson FI. 'Survey design and sampling' in de Cormack, D.F.S (Ed). The Research Process in Nursing (4th edn). Cornwall: Blackwell Publishing; 2000

[4] Bailey D, Research for the Professional: a practical guide (2nd edn). Philadelphia: Davis; 1997

[5] Beddoe A, Temperton D, Rafiqi A et al. (Implementation of IRR 99 and IR (ME) R 2000 in UK hospitals. Radiation Protection in Australia 2004; 21 (2): 59-70.

[6] Berger A. Magnetic Resonance Imaging, BMJ 2002; 32(4):35.

[7] Brink P and Wood MJ. Basic Steps in Planning Nursing Research (4th edn). Boston, MA: Jones and Bartlett; 1994

[8] Burns N and Grove SK. The Practice of Nursing Research: Conduct, Critique, and Utilization. Philadelphia: W.B Saunders; 2001.

[9] Burns N and Grove SK. The Practice of Nursing Research: Conduct, Critique, and Utilization (3rd edn). Philadelphia: W.B Saunders; 2003

[10] Buxton R and Liu T. Safety Guidelines for Conducting Magnetic Resonance Imaging (MRI). Experiments Involving Human Subjects, Version 3.1 Procedures: University of California,San Diego; 2007

[11] Byrns GE, Palatianos KH, Shands LA et al. Chemical hazards in radiology 2000; 15(2):203-208

[12] Carter DE. 'Descriptive research' in de Cormack, DFS (Ed). The Research Process in Nursing (4th edn). Oxford: Blackwell Science; 2000

[13] Center for Devices and Radiological Health.A primer on medical device interactions with magnetic resonance imaging systems, Food and Drugs Administration; 1997 Available at http://www.fda.gov/cdrh/ode/primerf6.html.Accessed: 31/12/2010.

[14] Center for Devices and Radiological Health.Guidance for industry and FDA staff, Criteria for significant risk investigations of magnetic resonance diagnostic devices, Food and Drugs Administration; 2003.Available at http://www.fda.gov/cdrh/ode/guidance.pdf.Accessed: 05/07/ 2010.

[15] Chaljub G. MRI safety for Healthcare Personnel. American journal of Roentgenology 2003; 11(4): 248-259.

[16] Clarke GD. MR Safety and Compatibility Issues at High Magnetic Fields, presentation for Utah Health Science Service Centre, San Antonio; 2007

[17] Colletti PM. Size H Oxygen Cylinder: Accidental MR Projectile at 1.5 Tesla. Journal of Magnetic Resonance Imaging 2004; 19: 141-143.

[18] Cornette LG, Tanner SF, Ramenghi LA et al. Magnetic Resonance Imaging, Fetal and Neonatal Education, 86:F171-F191; 2002. Available at www.archdischild.com.Accessed: 03/01/11.

[19] Dempsey MF, Condon B and Hadley DM. Ultrasound CT MRI. MRI Safety review 2002;. 23:392-401.

[20] Device Bulletin. Safety Guidelines for Magnetic Resonance Imaging Equipment in Clinical Use, Safeguarding Public Health: 03; 2007

[21] Emergency Care Research Institute. Patient death illustrates the importance of adhering to safety precautions in the magnetic resonance environments, Hazard report; 2001 Available at http//www.ecri.org/document/hazrd_MR1080601. Accessed: 15/10/2010.

[22] Elagin YP. The concept of safety, ensuring safety for infants undergoing Magnetic Resonance Imaging. Atomic Energy 1996; 80(6):389-393. Available at http// www.medscape.com/article/499273. Accessed 05/11/2010.

[23] Faulkner W. MRI: Basic Physics, Instrumentation and Quality Control. Maiden Mass: Blackwell Science; 2002

[24] Ferris JN, Kavnoudias H, Theil C et al. ACR Guidance Document for Safe MRPractices, American Journal of Roentgenology 2007; 188 (5): 1388-1394. Available at http:// www.acr.org/SecondaryMainMenuCategories/quality_safety/MRSafety/ safe_mr07.aspx.Accessed: 20/12/2010.

[25] Gilk T. MRI Suites: Safety outside the Bore. Patient Safety and Quality Healthcare, 1-8; 2006

[26] Gillies A. Using research in Nursing: a workbook for practitioners. Oxford: Radcliffe Medical Press; 2002

[27] Gould TA, How MRI works. Health, Willis HRH 08/09:1-8; 2008

[28] Gray DE, Doing Research in the real World. London: Sage; 2004

[29] Harding W. Concept of safety: Good Fellow. Texas: AFB; 2010

[30] Health Purchasing News. MRI accidents on the rise 2010; Available at http://findarticles.com/p/articles/mi m0BPC/is 10 29/ai n15685953.Accessed: 29/06/2010.

[31] Henner A and Servomaa A. The Safety Culture as a part of radiation protection in medical imaging, Proceedings of Third European IRPA Congress, Helsinki, Finland; 2010. Available at http://www.fda.gov/cdrh/ode/primerf6.html.Accessed: 31/12/2010.

[32] Johnson JN and Killion JB, Hazards in the radiology department. Entrepreneur: July-August; 2005; Available at http://www.entrepreneur.com/tradeljournals/article/ 134676835.html. Accessed: 20/01/11.

[33] Joseph NJ. MRI Safety for Healthcare Personnel.Online Radiology Continuing Education for Professional; 2006.Available at https://www.ceessentials.net/article7.html. Accessed: 09/08/2010.

[34] Junk R and Gilk T.Emergency Preparedness for Imaging Service Providers.Radiological Management 2005; 27(5): 16- 24.

[35] Kanal E, Barkovich AJ, Bell C et al. ACR Guidance Document for Safe MR Practices: American Journal of Roentgenology 2007; 188: 1447-1474.

[36] Kanal E and Shellock FG.Policies, Guidelines and Recommendations for MR Imaging safety and patient management and monitoring during MR examination.J MagnReson Imaging 1992; 2:247.

[37] Kanal E, Barkovich AJ, Gilk T et al. ACR White Paper on Magnetic Resonance (MR) Safety, presented as a combined paper to the American College of Radiology 2004; 1-24.

[38] Kanal E. Magnetic Resonance Safe Practice Guidelines of the University of Pittsburgh Medical Centre, summary from the Department of Radiology, University of Pittsburgh Medical Centre 2002; 155-163.

[39] Kanal E, Barkovich AJ, Gilk T et al. ACR Guidance Document for Safe MR Practices. American Journal of Radiology 2007; 188: 1-27.

[40] Kanal E, Borgstede JP, Barkovich AJ et al. White Paper on MRI Safety. American Journal of Roentgenology, Am J Roentgenol 2002; 178.1335-1347.

[41] Kangarlau A and Robitaille PM. Biological Effects and Health Implications in Magnetic Resonance Imaging.Concepts in Magnetic Resonance 2000;12 (5): 312-359.

[42] Kusumasuganda IGK. Introduction to MR Safety: Part I 2010; Available at http//mriforyou.bogspot.com.Accessed: 08/08/2010.

[43] Lahr W and Rowan R. How Metal Detectors Work. White's Electronics; 2004.

[44] Launders J, Preventing accidents and injuries in the MRI suite. The Joint Commission 2008; Issue 38: 1-3.

[45] Maiorova T, Stevens F, Van der zee J et al. Shortage in general practice deposit with feminisation of the medical workforce: A seeming paradox? A Cohort Study. BMC Health Service Research 2008, 8:262.

[46] Merchant SP, Halkett GK and Sale C. Australian radiation therapy: an overview. The Radiographer 2011; 58 (1):1-2.

[47] McRobbie DW, Moore EA, Graves M J and Prince MR. MRI, New York: Cambridge University Press; 2007

[48] Medical College of Wisconsin, MRI Safety Policies and procedures, Milwaukee, Wisconsin, 53226; 2009.

[49] New York Times, MRI accidents, 29 June 2010; Available at http://findarticles.com/p/
      articles/mi_m0BPC/is_10_29/ai_n15685953/ Accessed: 05/12/2010.

[50] Ordridge RJ, Fullerton G and Norris DG. MRI Safety limits: Is MRI safe or not? The
      British Journal of Radiology 2000; 73: 1-2.

[51] Phin D. Workplace Safety.General Safety Policies 2001; Available at http://
      www.inc.com/tools/2000/12/21572.html.Accessed: 07/01/2011.

[52] Polgar S and Thomas SA, Introduction to research in the Health Sciences (4thedn).
      London: Churchill Livingstone; 2000.

[53] Pontin D. 'Descriptive research' in de Cormack, DFS (Ed). The Research Process in
      Nursing (4thedn), Oxford: Blackwell Science; 2000.

[54] Price RP. The AAPM/RSNA physics tutorial for residents, MR imaging safety consid-
      erations 1999; 19:1641-1651.

[55] Sawyer-Glover A and Shellock FG. Pre-MRI Procedures Screening: Recommenda-
      tions and Safety Considerations for Biomedical Implants and Devices. Journal of
      Magnetic Resonance Imaging 2002; 12: 92-106.

[56] Schaefer DJ, Bourland JD and Nyenhuis JA. Review of patient safety in time-varying
      gradient fields. J. MagnReson Imaging 2000; 12:20-29.

[57] Schenck JF. Safety of strong, static magnetic fields. J MagnReson Imaging 2000;
      12:2-19.

[58] Shellock FG and Crues JV. MR Procedures: Biologic Effects, Safety, and Patient Care.
      Radiology 2004; 232 (3): 635 – 652.

[59] Shellock FG. MRI safety, bio effects and patient management 2002; Available at
      www.MRIsafety.com.Accessed: 29/08/2010.

[60] The Joint Commission, Preventing accidents and injuries in the MRI suite. Risk re-
      duction strategies 2008; Issue 38. Available at www.jointcommission.org/sentinele-
      vents/sentineleventalert/sea_38.htm Accessed: 29/06/10.

[61] University of California at San Francisco.MRI Safety Primer, Author, California;
      2011.

[62] Westbrook C, Talbot J and Roth CK. What do MRI radiographers really know? Mag-
      netic Resonance Imaging Journal 2009; 1(2): 52 -60.

Chapter 3

Spin Echo Magnetic Resonance Imaging

Mariluce Gonçalves Fonseca

Additional information is available at the end of the chapter

1. Introduction

Magnetic Resonance Imaging (MRI), as its name implies, is based on a magnetic resonance signal originating in the "spins" of hydrogen protons of a given patient's tissue undergoing magnetic resonance imaging under the action of a magnetic field [1].

Concerning the identification and characterization of tissues, the potential of MRI began to become apparent only in 1971, when it was realized that the magnetic relaxation properties of the nuclei differ among biological tissues. Furthermore, in the same tissue, this relaxation relied on the state of the vitality and integrity of tissues [2].

P. C. Lauterbur was the pioneer of imaging techniques for medical practice using MRI. In 1973, he described a method that produced a generation of a two-dimensional projection showing the density of the protons and the distribution of the relaxation times in a sample consisting of two water tubes. His studies were further improved by groups led by Hinshaw and Mansfield in England, Hutchinson in Scotland, Ernst in Switzerland, and Cho in Korea. Thus, alternative techniques have been developed to generate images that can assist both medical diagnoses and "in vivo" studies of biochemical reactions that occur at the cell level [1,3,4].

The most important factor for the formation of MRI is the "spin." In essence, the "spin" is a fundamental property of particles that make up the nucleus of the atom. Its concept was proposed by Samuel Abraham Goudsmit and George Eugene Uhlenbeck in 1925 [1].

Unlike the known images of Rx and CT, MRI does not use ionizing radiation but radiofrequency pulses.

The phenomenon of Magnetic Resonance Imaging manifests itself in molecular, atomic, electronic, and nuclear levels. In the latter case, its nature is magnetic, and therefore it is called nuclear magnetic resonance (NMR). It arises from the fact that certain nuclei possess an intrinsic angular moment referred to as "spin" and an associated magnetic moment. In medicine the term used is MRI. The term nuclear associated to it caused panic among patients, who believed the tests were harmful and painful to the tissues. In clinical trials, MRI is used to produce images of the body structures. This method has provided valuable assistance, since it is not invasive to biological tissues, and provides an excellent contrast between soft tissues [2,5,6].

2. MRI fundamental

In nuclei in which the "spin" protons are not paired, there is a resultant magnetic field which can be represented by a dipole magnetic vector. The magnitude of this field is called nuclear magnetic moment, and its existence causes the nuclei to respond actively to external magnetic fields. The nuclear magnetic vector does not remain static in one direction, but has a precessional motion or rotation around its axis (Figure 1).

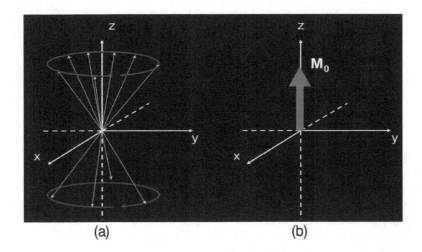

Figure 1. Schematic representation shows the spins in (A) the absence and (B) in the presence of an external magnetic field [3].

It is noted that in (A) without application of an external magnetic field, the protons are oriented in a random motion, while in (B) when placed in an external magnetic field B_0, the protons are aligned in the same direction, or in an opposite direction to the magnetic field. The slight preponderance of the spins in the same direction of the field creates a small resulting magnetization vector named M_0. This slight imbalance makes it possible to obtain images by RMI [3].

Two-thirds of the atoms that constitute the human body are hydrogen atoms, which contain only one proton in its nucleus. Therefore, they present a high-intensity magnetic vector, which increases their sensitivity to respond to external magnetic fields. In addition to hydrogen being the most abundant nucleus in biological tissues, its single proton results in more powerful magnetic moment than any other element. Due to these features, the hydrogen nucleus of biological tissues is the same one currently used to obtain the signal for the formation of images in MR procedures. However, other types of nuclei may be used to generate information on both the physiopathologic status and anatomy of tissues. Among other elements, we can cite carbon, oxygen, and sodium [7,8,9].

A radiofrequency pulse or excitation must be applied perpendicular to the main magnetic field in the frequency of precession or rotation of the hydrogen atoms (Larmor frequency) in order to obtain MR images. This radiofrequency pulse supplies energy to the resulting magnetization vector so that it is deflected to the transverse plane. Once the stimulation ceases, the magnetic vector returns to balance. This turning back to balance is measured and provides the generated resonance signal, which will be captured by the antennas of the MR apparatus [2,9].

3. Spin–echo sequence

In MRI, the most important pulse sequence is the "spin-echo" and its parameters are the repetition time (T_R) and echo time (T_E). Another important additional sequence is the "inversion-recovery" sequence, which promotes fat suppression, highlighting areas of injury with an additional parameter - the inversion time (T_I) [8,9,10].

Therefore, the keys to understanding MRI are physical principles, which include the magnetic properties of nuclei in biological tissues, the collective behavior of these biological tissues when excited by radio waves, and their relaxation properties, as well as the devices and techniques used to differentiate the tissues [7,9,10,11].

The technical parameters used to run a MRI were pulse sequences in "spin-echo" (SE) and " inversion-recovery " (Short T1 inversion STIR) to obtain images in T1 relaxation time (before and after injection of gadolinium contrast), in T2 relaxation time, and precontrast proton density (PD); Repetition time (TR), echo time (TE), and inversion time (TI); Section Plans (coronal or axial); Field of view (FOV), matrix size, number of acquisitions (NAQ), and number of sections, thickness, and interval between slices, and increment (F1), besides other functions to improve image quality [9,11].

The "spin-echo" pulse sequence [9,10,11] is used to obtain a signal by means of a 90° excitation pulse and a 180° inversion pulse, which were sent to the nuclei of hydrogen atoms of the tissues present in the region to be analyzed (Figure 2). These nuclei presented a rotating motion (precession), and when excited by a radio frequency coil (antenna), they start to rotate all at the same excitation frequency, resonating with each other. Once the stimulation is ceased, the MR signal is captured in form of signal or echo (Figure 3).

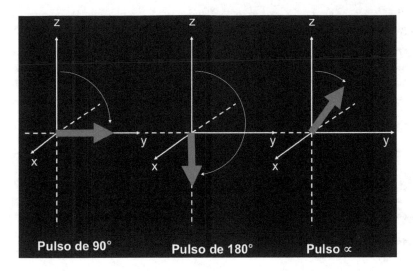

Figure 2. Radiofrequency pulse: 90º excitation pulse and a 180º inversion pulse, the pulse can be any value [3].

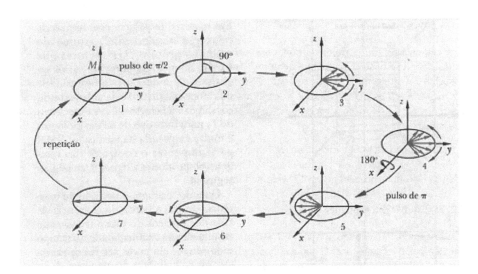

Figure 3. Illustration of the "spin-echo" (SE) imaging sequence [9,10].

When a pulse of 90º (π/2) is applied, the magnetization M initially in its equilibrium condition along the Z-axis (1) undergoes a 90º-displacement towards the y-direction (2). The tissues show a distribution of frequency of precession (3). There is a loss of coherence of the initial state (4). This loss can be reversed by applying a 180-degree pulse (π), which causes the spins of individual nuclei around the X-axis to rotate 180 degrees (5), rephasing (6) and regenerating the signal, referred to as spin-echo (7).

The 90º pulse plus the 180º pulse produced an echo, which is repeated several times during the study in the analyzed region. This echo is referred to as the repetition time (T_R). The echo time (T_E) is the duration between the middle of a 90º pulse and the middle of an echo (Figure 4).

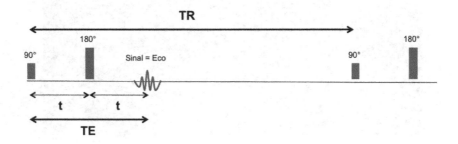

Figure 4. SE pulse of 90º and applied time (TE/2) of pulse RF of 180º [3].

3.1. Conventional spin–echo sequence

The sequences of pulses in conventional spin-echo can be used in almost all tests. T1-weighted images are useful to demonstrate anatomy, but they can also demonstrate diseases when associated with contrast enhancement. T2-weighted images also demonstrated diseases. Tissues affected by diseases appear edematous and/or vascularized. They have higher water content and therefore, a strong signal on T2-weighted images. Thus, they can be easily identified.

Usually, in conventional spin-echo sequence a short T_R a short T_E will give a T1-weighted image, a long T_R and short T_E (first echo) will give a proton density image, and a long T_R and long T_E (second echo) will give a T2-weighted image [10].

3.2. Fast spin–echo sequence

The fast spin-echo sequence is a spin-echo sequence, but with the time of the exam dramatically shorter than the conventional spin-echo. To understand how rapid the fast spin-echo sequence is, we should review how data is obtained in the conventional spin-echo. A 90º excitation pulse is followed by a 180º rephasing pulse. Only one encoding phase step is applied by T_R in each section and just one K-space line is completed by T_R [10,12,13].

Generally, the contrast observed in fast spin-echo images is similar to that of the conventional spin-echo images. Therefore, these sequences are useful in many clinical applications. In the central nervous system, pelvis, and musculoskeletal regions, the fast spin-echo sequence has practically substituted the conventional spin-echo. In the chest and abdomen, however, the respiratory artifacts are sometimes problematic in cases where the respiratory compensation techniques are not compatible with the programs fast spin-echo, which is counterbalanced to some extent by the fact that shorter examination times in fast spin-echo sequence enable the production of images with fewer respiratory artifacts in [9,10,11,13,14,15].

There are two differences in contrast between the pulse sequence of the conventional spin-echo and fast spin-echo, both of which are due to the 180º pulse repeated at short intervals following the sequence of echoes. First, the adipose tissue remains clear on T2-weighted images due to multiple RF pulses that reduce the effects of spin-spin interactions in this tissue. However, the fat saturation techniques may be used to compensate for this. Second, the 180º repeated pulses may increase the magnetization transfer, so that the muscles appear darker on the fast spin-echo images than on the conventional spin-echo images. Additionally, multiple 180º pulses reduce the effects of magnetic susceptibility, which may be detrimental when looking for small haemorrhages [10].

The advantages of fast spin sequence are that metal implant artifacts are significantly reduced in rapid sequences.

In fast spin-echo T1-weighted images, effective TE and TR are short; on T2-weighted effective TE and TR are long TR; on proton density weighting/T2-weighted images, effective TE is short and effective TR is long [10,11,13,15].

The advantages are: Greatly reduced examination times, better image quality, and more information on T2-weighted images. We can use high-resolution matrices and multiple numbers of excitations (NEX). However, some effects of increased flow and movement are incompatible with some options of image acquisition, such as fat tissue bright on T2-weighted images, blurred images can occur because data were collected at different TE time, decreased magnetic susceptibility effect, because multiple 180º pulses produce excellent returning phase, so that one must not use it in case of suspected bleeding [4, 9,10,13,14,15].

The "inversion-recovery" sequence is used to promote suppression or fat saturation, high-lighting areas of injury. The process was the reverse of the "spin-echo" sequence. There was an inversion followed by a recovery by applying 180º inversion pulses, which inverted the spins of the fatty tissue region examined by 180º, followed by 90º recovery pulse. Subsequently, a 180º repolarizing pulse was applied to produce a spin-echo. In this sequence, the repetition time (T_R) is the time between each 180º pulse. The inversion time (T_I) is the length of time the fat (spins) took to recover from this complete inversion (Figure 5).

This process allowed the fat to become dark or hypointense, differing itself from the lesions. This happened because the inversion of its spins caused a total loss of energy/magnetization. Consequently, there is no sign for it [10].

The field of view (FOV) determines the size of the anatomy covered during the selection of the tissue section to be analyzed either in a coronal or axial plane.The forming unit of a digital image is the pixel. The brightness of each pixel represents the power of the MR signal produced by a volumetric imaging of the patient or volumetric pixel or Volumetric Picture Element (voxel). The voxel is a volume element representing the tissue inside the patient. It is deter-

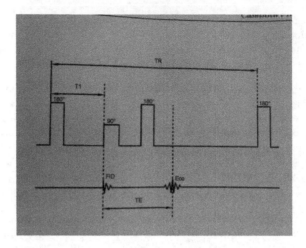

Figure 5. Illustration of the resonance image inversion-recovery pulse sequence. A 180° pulse inversion is applied followed by a 90° recovery pulse, as well as a 180° repolarization pulse. T_R, T_E and T_I are also shown [16].

mined by the pixel area and the thickness of the section. Thus, the size of the matrix is determined by the number of pixels of the anatomy covered during the selection of the tissue section to be analyzed. This size is indicated by two values. The first one corresponds to the number of frequencies sampled and the second to the number of phase codings performed [7,10,13].

Frequency codification is the reading of a signal along the longest axis of the anatomy. The phase codification is the reading of a signal along the short axis of the anatomy. Thus, a matrix size of 256 x 192 indicates that 256 encoding frequencies and 192 encoding phases are performed during a sequence [9,10].

The number of acquisitions (NAQ) represents the number of times that data are acquired within/into the same pulse sequence [10,11].

The number, thickness and intervals of the sections are defined according to the type of lesion. Other functions are used to improve image quality. Its use allows viewing only the sections selected [10,11].

4. Tissue parameters

The images primarily reflect the distribution of free hydrogen nucleus and the way it responds to an external magnetic field. Thus, this response determines different relaxation times known as T1 and T2. The pathological processes cause relaxation time to change in relation to the tissues of the nervous and musculoskeletal system, and the signal intensity is reflected [7,9,16].

4.1. Tissue relaxation time T1

Required for recovery of about 63% of the magnetization along the longitudinal direction after a 90º pulse are generally more anatomical, since the fat planes are hyperintense, perfectly delimiting muscle planes and vascular structures. When paramagnetic agents (contrast) are associated, they demonstrate the skin changes with much more specificity. It is used to evaluate the anatomic structures of the injured limb in MRI and SE sequences before and after contrast. The mechanism is based on the application of a 90º RF pulse that diverted the longitudinal magnetization towards the transverse plane. Subsequently, there is a recovery of this energy diverted to the initial longitudinal axis. In a more simplified way, T_1 is the time required for the initial 63% recovery of the magnetization along the longitudinal axis after the application of 90-degree RF pulse (Figures 6 &7) [7,9,10].

Thus, the signal intensity (brightness) emitted by the tissues depends solely on its ability to recover the magnetization faster or slower after the application of a 90-degree RF pulse.

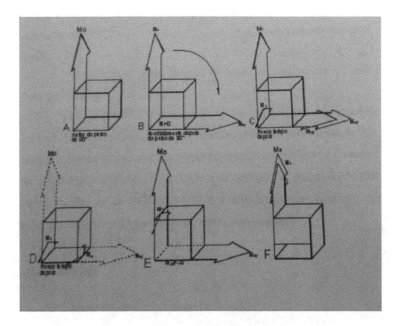

Figure 6. Schematic representation of T_1 relaxation time.

Note that the relaxation time T1 begins in (A) before the 90º pulse when the magnetization M_0 is in the axis. Just after the 90º pulse, the magnetization is zero and the transverse is maximum (B). A short time later, there is the recovery of the resulting longitudinal magnetization (C) representing the start of recovery T1 (D, E), and in (F) occurs the 63% recovery of the initial magnetization [16].

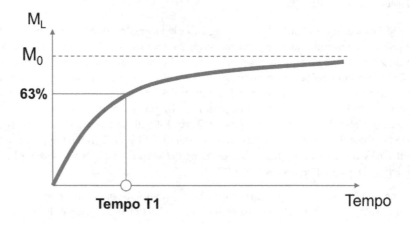

Figure 7. Relaxation time T1: recovery 63% of the magnetization along the longitudinal direction after a 90° pulse [3].

4.2. Tissue relaxation time T2

Tissue relaxation time T_2 is used throughout the SE sequence to detect lesions. At T_2 time, there is a magnetization shift or loss. The tissues' capacity to lose magnetization faster or slower is what determines the signal strength. T_2 time is the time required for the transverse magnetization to drop up to 37% of its initial value after the application of a 90-degree pulse (Figure 8 & Figure 9) [7,9,10].

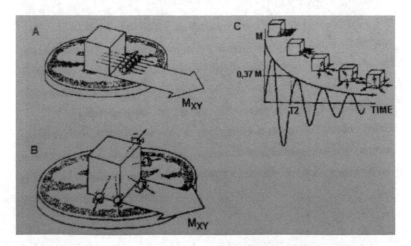

Figure 8. Schematic representation of T_2 relaxation time.

In (A) are representative protons of a tissue section. Soon after a 90-degree pulse, the protons are on the same transverse plane and in phase with each other. Their magnetic vectors all point in the same direction. (B) After a very short period of time, these protons are out of phase, and their magnetic vectors are pointing to different directions. This decreases the power of the transverse magnetization vector Mxy. (C) T2 is shown as the time interval required for the transverse magnetization drops to 37% of its original value [16].

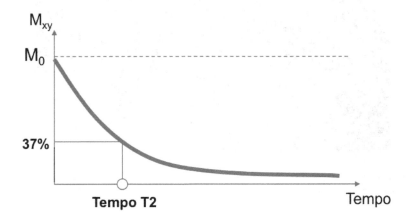

Figure 9. T2 shown as the time interval required for the transverse magnetization drops to 37% of its original value [3].

5. Contrast

The contrast agent used is a paramagnetic metal called gadolinium (GDL). It is associated with a water-soluble component diethylenetriaminepentaacetic acid (DTPA) that acts on the damaged tissues facilitating their identification [17, 18].

It is administered intravenously at a dose of 0.2 mL/kg on T1-weighted images through section planes determined according to the location and type of injury [17,18].

Patients who receive contrast are asked to abstain from all food and liquid for two hours in order to avoid adverse effects [17,18].

Local lesions are studied for the presence or absence, type, and thickness of the damaged tissues. The determination of the type of lesion is accomplished through changing the signal presented by damaged tissues in relation to normal tissue. The classification of injured tissues into hypointense or hyperintense, depends on the signal intensity (darker or lighter) visualized on the images during the screenings and on an expert testimony (Figure 10) [17,18].

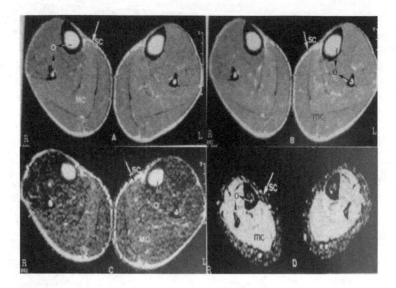

Figure 10. Normal tissue in MRI in axial sections in the "spin echo" sequence taken from the lower limbs (calf) in T_1 pre (A) and (B) post-contrast injection, T_2 relaxation times (C) and inversion-recovery" sequence (D) used to promote suppression or fat saturation [16].

In these images, the tissues present themselves with their normal callibre vascular structures and anatomic topography, as well as their musculature with preserved sign and normal morphological aspect. The images also present the bone structure of their cortical portions and characteristic medullar signal, and preserved anatomical aspect [16].

For images of the central nervous system, "Figure 11" illustrates the characteristics in normal tissue relaxation time T1 before and after contrasts, which are used to differentiate normal tissue from the pathological ones [19,20].

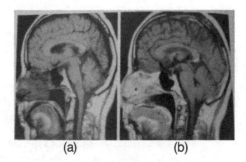

Figure 11. Image of a normal central nervous system (sagittal plane) on pre-contrast (A) and post-contrast (B) sequences spin-echo T1-weighted images.

Note all structures with normal anatomic aspects with enhancement in sequence with contrast, indicated by arrows [21].

6. MRI machine

A magnetic resonance imaging (MRI) machine consists of a main magnet that provides a closed or open scanning system. It is a permanent superconductor. Its power field ranges from 0.23, 0.5, 1.0, 1.5 up to 3.0 Tesla total power field. Internally, the main magnet is composed of homogenizing coils, gradient coils, and radiofrequency (RF) transmitter and receiver coils. These may be located internal or external to the main magnet. The function of these components is to capture the signal or echo generated by the tissues (tissue parameters) when in contact with the magnetic field and technical parameters used [9,10,12]. The machine also comprises computers and image processors, which make it possible to acquire and visualize the image on the operator's console monitor (Figure 12 & Figure 13).

The technical parameters are those dependent on the device and set up by the operator based on examination protocols.

Initially, the patients are placed on the examining bed. The region (lesion) being examined is highlighted by a source of light directed and positioned in the center of the magnet. Afterwards, the device is set up with a specific test protocol according to the limb damaged. Following, we made a first localization sequence in the desired section plane. Thus, we could design other section planes from the image formed [10,12].

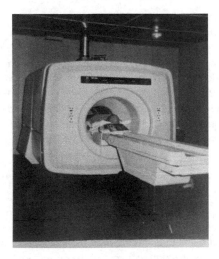

Figure 12. Closed field magnetic resonance imaging machine [16].

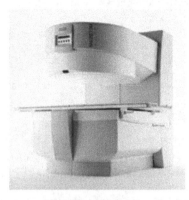

Figure 13. Open field magnetic resonance imaging machine [9].

The physical principles of the open field MRI are the same as that of the closed field MRI, which uses a strong magnetic field created by the movement of electrical currents within a series of spiral coils located inside the machine [7,9].

The open field MRI is a breakthrough technology to obtain images of the human body without constraints for patients with claustrophobia (fear of closed spaces), obesity, as well as children and elderly people [7,9,12].

The advantages of the open-field MRI are associated to a machine having large side openings that allows the patient to be examined with more tranquillity, comfort, and convenience. It also helps to obtain a better quality of the images [7].

In practical terms, we can consider the MRI machine as a large and powerful magnet. The acquisition of spin-echo images can be understood as follows: The patient is placed into the MRI machine. Once inside the machine all hydrogen ions in the different body tissues will align parallel with the magnetic field of the machine. Then, a coil emits RF pulses that cause the axis of these ions to change 90º. When the coil turns off, the ions tend to realign with the magnetic field, but with different intensities and speeds according to the type of tissue in which they are found. This difference in intensity and time is captured and quantified by the device that locates and defines shades of grey for each point detected. The information is processed by a computer workstation that accomplishes the construction of images in the frontal, sagittal, and axial planes [10,12].

The technical parameters are those dependent on the device and set up by the operator based on examination protocols.

Initially, the patients are placed on the examining bed. The region (lesion) being examinedis highlighted by a source of light directed and positioned in the center of the magnet. After-wards, the device was set up with a specific test protocol according to the limb damaged. Following, we made a first localization sequence in the desired section plane. Thus, we could design other section planes from the image formed [16,19].

The obtained images are recorded and photographed on film (Figure 14). The final appearance will depend not only on intrinsic properties of tissues but also on technical aspects such as pulse sequences or time factors that are chosen and machine quality.

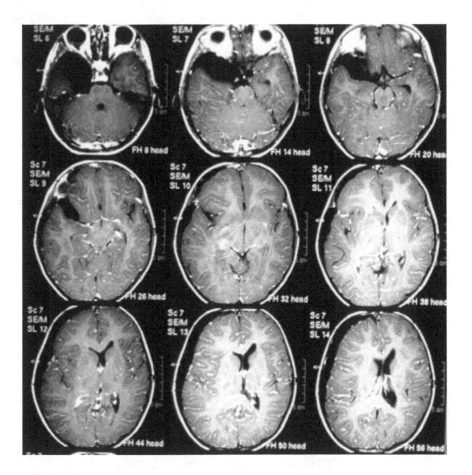

Figure 14. MRI obtained in SE sequence in the axial plane of the skull [19].

For each type of exam of any region of the human body, there is a specific protocol to obtain MR images, most are used for detecting soft-tissue lesions of the structures that make up the central nervous system and skeletal muscle.

7. Examples of MRI protocols and applications by SE sequence

This method has been widely used in the diagnosis of diseases located in the structures of the nervous and musculoskeletal systems. Thus, MRI is an imaging method that provides excellent contrast between soft tissues, due to its high spatial resolution. Therefore, from the anatomical point of view, MRI is the best choice for evaluation of the structures that make up the musculoskeletal system. The protocols on Table 1 and Table 2 were used to acquire the images of the following images which represents examples of very interesting applications of MRI.

Section planes		Cor loc	AXT1	AXT2	AX T1 GDL	Cor T1 GDL
FOV		SE42	SE30 IR 25	SE30	SE 30	SE 35 IR 35
TR in ms		SE30	SE850 IR2000	SE 2000	SE 850	SE 750 IR2000
TE in ms		SE25	SE25 IR 90	SE 40	SE 25	SE 25 IR 90
TE(2°) in ms		-	-	SE80	-	-
TI in ms			IR 25			IR 25
Interval		SE15	SE10 IR 12	SE10	SE 10	SE 10 IR 10
N^{umber} of sections		SE 6	SE11 IR 12	SE 12	SE 11	SE 12 IR 12
Thickness in Mm		SE10	SE 5 IR 5	SE 5	SE 5	SE 5 IR 5
NAQ		SE 1	SE 4 IR 1	SE 2	SE 4	SE 4 IR 1
Matrix	SE IR IR	192x192	256x192 224x256 256x192	256x192	256x192	192x256 256x256
(F1)		-	SE10 IR 11	SE 10	SE 10	SE 8 IR 11

Table 1. Exam protocol and values of technical parameters and tissue for evaluation of lesions in the lower limb (0.5 Tesla MRI). Body and head coils.

Section planes	AX LOC	COR T1	AX T2	AX T1	AX T1 GDL
FOV	25	15	25	22	22
TR in ms	SE 320	SE 750	SE 2000	SE 750	SE 750
TE in ms	25	30	40	25	25
TE(2°) in ms	-	-	80	-	-
Interval	7	5	5	8	8
N^{umber} of sections	4	12	13	12	12
Thickness in mm	5	5	5	5	5
NAQ	1	2	2	4	4
Matrix	192x192	192x192	256x192	192x160	192x160
(F1)	-	-	10-8	-	-

Table 2. Exam protocol and values of technical parameters and tissue for evaluation of upper limb injuries (0.5 Tesla MRI). Elbow in shoulder coil.

7.1. Application to musculoskeletal tissue lesions

The MR images on the axial plane (AX) show the skeletal muscle and central nervous system. In the sequence, lesions diagnosed as edema and blood in subcutaneous, perimuscular, and muscular tissues and central nervous system structures in pre- and post-contrast T1 and T2 times (Figures 15, 16 &17). Edema presents as a hypointense signal on pre-contrast T1 time and enhanced on pre-contrast T1 time and hyperintense on T2 time. Lesions identified as haemorrhagic lesions present a hypersignal on pre- and post-contrast T1 and T2 times [21,22,23].

The edema corresponds to an increase of water content into the extracellular space and/or into the intracellular compartment. T2-weighted sequences are the main time interval that detects this increase in the form of an intense area of hypersignal in [21,22,23].

In haemorrhagic lesions or in the presence of degradation components of blood in any tissue often give the hyperintense signal on T1 and T2. They are a consequence of a local vascular injury [22,23].

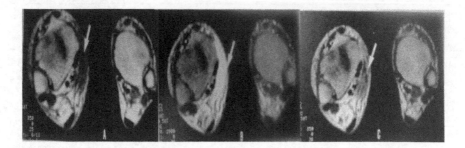

Figure 15. MRI of the right foot showing edema in subcutaneous tissue characterized by (A) hyposignal on T1 (B) hyperintense on T2, and (C) enhanced on post-contrast T1. Musculature and perimuscular areas preserved [16].

Tissue lesion and inflammatory processes related to the musculoskeletal system cause changes in the relaxation times T1 and T2 and reflects the signal intensity. The inflammatory processes increase the signal intensity on T2-weighted images and the swelling causes an increase of water in the tissues that determines the signal changes observed [22].

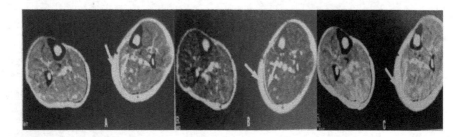

Figure 16. MRI showing the left calf. The injury is consistent with subcutaneous tissue and perimusculare region mild haemorrhage characterized by (A) isointese to hyperintense signal on T1, (B) hyperintense signal on T2, and (C) enhanced on post-contrast T1. The presence of blood in the perimuscular region is well visualized on relaxation time T2.

Bleeding observed in subcutaneous and muscle tissues is generally different from that resulting from the degradation process known in the pathologies of the central nervous system. In these pathologies, the bleeding is presented in various stages of degradation and is known as oxyhemoglobin and/or deoxyhemoglobin, (intracellular or free) methemoglobin, and hemosiderin. Thus, these various stages interfere with the lesion signal intensity and stage interpretation [24,25].

As to the skeletal muscle, it may present in the form from an iso to hyperintense signal at all relaxation times before and after contrast injection [16].

It is noted that in these images the edema in association with haemorrhage usually presents themselves with signal hyperintensity on the T2-weighted images.

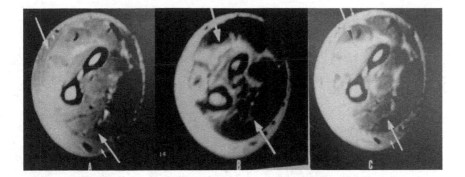

Figure 17. MRI of the right forearm indicating extravasation of blood into muscle tissue characterized by (A) isointense to hyperintense signal on T1-weighted image (B) hyperintense signal on T2-weighted image (C) enhanced on post-contrast T1-weighted image [23].

7.2. Tumor injuries detected in the central nervous system

The vast majority of intracranial tumors present a high-protein density, a long T1 and T2, so generally there is a hypo signal on T1-weighted (short TE-TR) and a hyperintense signal on T2-weighted sequences (long TE-TR). Thus, the signal variations are not very specific (Figure 18 & 19). The application presented in Figure 18 an Figure 19 concerns the examination of rectal adenocarcinoma and meningioma of left ventricle fibrous trigonum respectively.

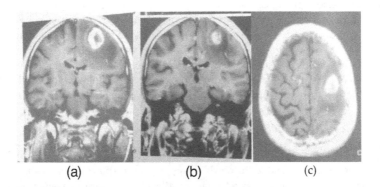

(a)   (b)   (c)

Figure 18. A and B are frontal section images on T1-weighted imaging. C After contrast injection. The hyperintense tumor (A, B, C) gives the perfect location of both the metastasis and the hypointense perilesional edematous reactions [21].

Whatever the sequence used after contrast injection, the parenchymatous reaction edema is visualized with hypointense signal on T1 pre- and post-contrast (A, B) and with hyperintense signal on T2 (C, D). Note the displacement to the right of the median structures of the septum pellucidum.

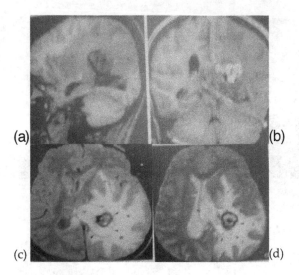

Figure 19. T1-weighted imaging sequences in sagittal plane (A) and T2-weighted imaging sequence in axial plane (C, D) after contrast injection on T1-weighted sequence in frontal plane (B) [21]

Cerebral edema can be of three types: vasogenic corresponding to a disruption of the blood-brain barrier to the passage of a protein-rich filtrate in the brain extracellular spaces, nonspecific outcome of multiple pathological processes (primary tumors, metastases, haemorrhage, trauma, inflammatory processes and infection). It manifests as a hyperintense signal area of white matter, respecting the gray matter. The accomplishment of a sequence with strong T2-weighted can evidence that it is due to the edema's persistent hyperintense signal in contrast to the tumour's decreasing signal. However, the sequences on T1 post-contrast are the ones bounding the lesion; the earliest manifestation form of infarction is the cytotoxic edema. The ischemia leads to an early failure of the membrane pump, which allows water and sodium to enter the cells. It presents itself as a hyperintense signal involving the white and gray matter [21,24,25,26].

Interstitial edema is found in hydrocephalus with passage of transependymal water into the brain tissue from the ventricular cavities, essentially around the lateral ventricles [21].

The water being highly bound to the neighboring proteins displays a significant decrease of T1. The interstitial edema can be viewed paradoxically under the form of a hyperintense signal on T1-weighted sequences, while still naturally with hyperintense signal on T2-weighted sequences [21,27,28].

Thus, the contrast injection increases the specificity in the detection of lesions. The paramagnetic agents such as the gadolinium (GDL) associated with a chelating agent - diethylenetriamine pentaacetic acid (DTPA) - is a safety water soluble. After its application, around 80% is excreted by the kidneys in three hours, and the remaining is recovered in stools and eliminated within a week [18].

The MRI scan is the method of choice for the evaluation of tumors. The sequence systematic practice, mainly of spin echo sequences in different space planes (particularly in axial and sagittal planes), and the intravenous injection of GDL allows a perfect assessment of the tumours [21,27,30].

8. MRI and artifacts

The quality of MR images depends on multiple factors that can significantly alter the outcome of the tests and therefore, the diagnosis of lesions. The so-called artifacts can determine impairment in the image formation and may be inherent to the method (apparatus, pulse sequence) and those related to the patient (involuntary physiologic recurrent movements and involuntary non recurrent movements). The physiological recurrent movements are related to breathing and heartbeat, while involuntary non periodic can be determined by swallowing or spontaneous movements of patients. The artifacts generally can alter the quality of the image during its acquisition. Therefore, in some cases, they interfere with the interpretation of the diagnosis [21,31].

9. MRI scanning: Risks and contraindications

Up to 2.5 Tesla, the magnetic field does not trigger any biological or genetic risk.

The risks and contraindications for MRI are very rare, but they should be known to avoid an accident or scheduling of an unnecessary exam.

Risk factors are associated to a magnetic field that can produce heat, suffocation in case of discharge of a supra-conductor magnet with brutal gasification of the fluids that cools the magnet, patients' local burns caused exceptionally by the twisting of the antenna surface wire or its deterioration by the "coil" effect [21,32,33].

The exam is contraindicated for patients with cardiac pacemakers that can be affected temporarily or permanently with risk of heart failure or rhythm disturbances; these risks exist regardless of the intensity of the magnetic field, metal and ferromagnetic bodies, and pregnant women [32,33].

10. Conclusions

Studies in MRI to diagnose soft-tissue injuries, mainly of the skeletal muscle and central nervous system, indicated that the most-used pulse sequence is the spin echo. Through this sequence it is possible to obtain images in axial, frontal, and sagittal planes. According to these studies, the images obtained in the axial plane are those that show the lesions in detail.

The sequences with contrast images obtained on T1-weighted images are the most important to determine areas of injury with greater specificity. T2-weighted images allow accurately diagnosed injuries. Paramagnetic agents are of primary importance and its use in MRI provides information about the behavior of the lesions.

MRI scans can be conducted in all regions of the body such as brain, spine, joints (shoulder, knee), extremities, chest, abdomen, and others. It is an excellent method for detecting tumours and other soft-tissue lesions based on the criteria of patient safety in relation to the magnetic field, pathology and site to investigate, as well as technical parameters and tissue, which are critical in image acquisition.

Nomenclature (list of symbol)

The nomenclature represents the protocols used to acquire the images of tissues in MR spin echo sequence of skeletal muscle and central nervous system.

AX LOC. Axial section plane locate

COR LOC. Coronal section plane locate

COR T1. Coronal section plane tissue relaxation time T1

AX T2. Axial section plane tissue relaxation time T2

AX T1. Axial section plane tissue relaxation time TI pre-contrast

AX T1 GDL. Axial section plane tissue relaxation time T1 pos-contrast

GDL. Contrast agent paramagnetic metal (gadolinium)

SE. Spin Echo sequence

IR. Inversion-recovery sequence

FOV. Field of view determine the size of the anatomy covered during the selection of the tissue section

TR. Repetition time

TE. Echo time

TE(2º). Two sequences in echo time

TI. Inversion time

Interval. Interval between slices to image quality

Number of sections. Number of slices to image quality

Thickness. Thikness of slices image quality

NAQ. number of acquisitions represents the number of times that data are acquired within/into the same pulse sequence

Matrix. Codification frequency and phase codification along the longest and short axis of the anatomy

F1. Increment to image quality

Acknowledgements

I would to thank Dr. José Ricardo de Arruda Miranda, Dr. Benedito Barraviera, Dr. José Morceli, Dr. Seizo Yamashita, Ms. Maria Rita de Cássia Mathias, Alexandre Lins Werneck, PH.D and Prof. David Mercer for their help and assistance.

Author details

Mariluce Gonçalves Fonseca

Federal University of Piaui, School of Medicine, UNESP, Botucatu, Brazil

References

[1] Crooks LE, Ortendahl DA, Kaufman L, Hoenninger J, Arakawa M, Watts J, Cannon C, Brant-Zawadzki M, Davis PL, Margulis RA. Clinical efficiency of nuclear magnetic resonance imaging. Radiology 1983;146 123-128.

[2] Lauterbur PC. Image formation by induced local interactions: examples employing nuclear magnetic resonance. Nature 1973;242 190-191.

[3] Mazzola AA. Magnetic resonance: principles of image formation and application in functional imaging. Revista Brasileira de Física Médica 2009;3(1) 117-118.

[4] Foster MA. Magnetic resonance in medicine and biology. New York: Pergamon Press; 1984.

[5] Moonen CT, van Zijl PC, Frank JA, Le Bihan D, Becker ED. Functional Magnetic Resonance Imaging in Medicine and Physiology. Science 1990;250(4977) 53-61.

[6] Bloch F. Nuclear induction. Physical Review 1946;70(7-8) 460-474.

[7] Buhong S. Magnetic resonance imaging. Physical and Biological Principles. St Louis: Moseby; 1996.

[8]   Purcell EM, Torrey HC, Pound RV. Resonance absorption by nuclear magnetic moments in a solid. Physical Review 1946;69 37-38.

[9]   Bernstein MA, King KE, Xiaohong JZ. Handbook of MRI pulse sequences. London: Elsevier; 2004.

[10] Hahn EL. Spin echoes. Physical Review 1950;80(4) 580-594.

[11] Mansfield P. Multi-planar imaging formation using NMR spin echoes. Journal of Physics 1977;10 155-158.

[12] Westbrook C. Handbook of MRI Technique. Oxford: Blackwell Science; 1994.

[13] Werhli F. Fast scan magnetic resonance – principles and applications. New York: Raven Press; 1991.

[14] Claasen-Vujcic T, Borsboom HM, Gaykema HJ, Mehlkopf T. Transverse low-field RF coils in MRI. Magnetic Resonance in Medicine 1996;36(1) 111-116

[15] Henning J, Nauerth A, Friedburg H. RARE imaging: a fast imaging method for clinical MR. Magnetic Resonance Medicine 1986;3(6) 823-833.

[16] Fonseca, MG. Lesion tissues in patients snake bite of genus *Bothrops* and *Crotalus*: Clinical, laboratory study and evaluation by Magnetic Resonance Imaging, PhD thesis. School of Medicine UNESP Botucatu Brazil; 2000.

[17] Bloem JL, Reiser MF, Vanel D. Magnetic resonance contrast agents in the evaluation of the muscle-skeletal system. Magnetic Resonance Quarterly 1990; 6(2) 136-163.

[18] Mathur-De Vré R, Lemort M. Invited review: biophysical properties and clinical applications of magnetic resonance imaging contrast agents. British Journal of Radiology 1995; 68(807) 225-247.

[19] Amaro Jr E, Barker GJ. Study design in MRI: basic principles. Brain and Cogn 2006;60(3) 220-232.

[20] Kwong KK, Belliveau JW, Chesler DA, Goldberg IE, Weisskoff RM, Poncelet BP. Dynamic magnetic resonance imaging of human brain activity during primary sensory stimulation. Proceedings of the National Academy of Sciences USA 1992;89(12)5 675-679.

[21] Doyon D, Cabanis EA, Iba-Zizen MT, Laval-Jeantet M, Frija J, Pariente D, Idy-Peretti I. IRM Imagerie par résonance magnétique. Masson; 1997.

[22] Fonseca MG, Mathias MRC, Yamashita S, Morceli J, Barraviera B. Local edema and hemorrhage caused by *Crotalus durissus terrificus* envenoming evaluated by magnetic resonance imaging (MRI). Journal of Venomous Animals and Toxins 2002;8 49- 59.

[23] Fonseca MG, Mathias MRC, Yamashita S, Morceli J, Barraviera B. Tissue damage caused by *Bothrops* sp envenoming evaluated by magnetic resonance imaging (MRI). Journal of Venomous Animals and Toxins 2002;8 102–111.

[24] Pauling L, Coryell CD. The magnetic properties and structure of hemoglobin, oxyhemoglobin and carbonmonoxyhemoglobin. Proceedings of the National Academy of Sciences USA 1930;22 210-215.

[25] Thulborn KR, Waterton JC, Matthews PM, Radda GK. Oxygenation dependence of the transverse relaxation time of water protons in whole blood at high field. Biochimica et Biophysica Acta 1982;714(2) 265-270.

[26] Buxton RB, Wong EC, Frank LR. Dynamics of blood flow and oxygenation changes during brain activation: the ballon model. Magnetic Resonance Medicine 1998;39(6) 855-864.

[27] Sunaert S. Presurgical planning for tumor resectioning. Journal Magnetic Resonance Imaging 2006;23(6) 887-905.

[28] Ogawa S, Lee TM, Kay AR, Tank DW. Brain magnetic resonance imaging with contrast dependent on blood oxygenation. Proceedings of the National Academy of Sciences USA 1990;87 9868-9872.

[29] Ogawa S, Tank DW, Menon R, Ellermann JM, Kim SG, Merkle H, et al. Intrinsic signal changes accompanying sensory stimulation: functional brain mapping with magnetic resonance imaging. Proceedings of the National Academy of Sciences USA 1992;89(13) 5951-5955.

[30] Bandettini PA, Wong EC, Hinks RS, Tikofsky RS, Hyde JS. Time course EPI of human brain function during task activation. Magnetic Resonance in Medicine 1992;25(2) 390-397.

[31] Shellock FG, Shellock VJ. Vascular acces ports and catheters: ex vivo testing of ferromagnetism, heating, and artifacts associated with MR imaging. Magnetic Resonance Imaging 1996;14(4) 443-447.

[32] Shellock FG, Kanal E. Guidelines and recommendations for MR imaging safety and patient management. Journal of Magnenetic Resonance Imaging 1994;1(1) 97-101.

[33] Shellock FG, Schaefer DJ, Gordon CJ. Effect of a 1,5 T static magnetic field on body temperature of man. Magnetic Resonance in Medicine 1986; 3 644-647.

Chapter 4

Mathematics and Physics of Computed Tomography (CT): Demonstrations and Practical Examples

Faycal Kharfi

Additional information is available at the end of the chapter

1. Introduction

The visible light being reflected by the majority of the objects which surround us, we can apprehend our environment only by the properties of the surface of the objects which compose it. To exceed this limit and to explore the intimacy of the matter, a dedicated imaging techniques and instruments, which used penetrating radiations like X-rays, neutrons, gamma, or certain electromagnetic or acoustic rays to explore internal structure, are developed. The tomography is one of these developed techniques that allow 2D and 3D interior object examination. By combination of a set of measures and thanks to computational and images reconstruction methods, it provides cartography of attenuation parameters characteristic of the radiation/object interaction, according to one or more transversal plans (slices). It thus makes possible to see on TV monitor the interior of bodies and objects, whereas before one had access either by pure imagination, by interpreting indirect measurements, or by cutting out the objects materially. In the case of the medical imaging, for example, this direct observation requires a surgical operation. This formidable invention thus enables us to discover the interior of human bodies and different objects which surround us and their organization in space and time, without destroying them.

The tomography thus constitutes an instrument privileged to analyze and characterize matter, that it is inert or alive, static or dynamic, of microscopic or macroscopic scale. While giving access to the structure and the shape of the components, it makes possible the apprehension of the complexity of the objects studied. The computed tomography is a technique of acquisition of digital images (projections). It generates a computer coding of a digital representation of an area of interest through a patient, a structure or an object. The tomography thus provides a virtual representation of reality, in the intimacy of its composition. This numerical coding then will facilitate the exploitation, the exchanges and the information storage associated. It becomes thus possible by appropriate processing to detect the presence of defects, to identify the internal structures and to study their form and their position, to quantify the variations of density, to model the internal components and to guide the instruments of intervention in medicine. Finally, the user will be also able to take benefit from a large variety of existent software and algorithms for the tomography digital images processing, analysis and visualization.

A CT system gathers several technological components. Its development requires the participation of the end users - as the doctors, the physicists or the biologists - to specify the needs, of the engineers and researchers to develop the novel methods and, finally, of the industrial teams to develop, produce and market these systems.

In this work, we are interested in the physics and mathematics related to the main phases of computed tomography, namely: the scanning or projection phase and the phase of 2D or 3D image reconstruction by various analytical methods such filtered back-projection method (FBP). In this context, all the mathematical equations and relations which seemed ambiguous or not clear are explained and mathematically demonstrated with some illustrative examples. This chapter is organized in the form of a main body text with sub-sections presenting a brief explanation of most important concept or the detailed demonstration of any equation or expression, and this, each time that seemed necessary. The algebraic methods for image reconstruction in tomography will be also outlined.

This chapter is intended as an introduction to Computed Tomography. It was written not only for those persons who have some familiarity with other imaging techniques such radiation transmission radiography but also for novices in the field of digital imaging. The chapter begins with some simple, yet fundamental, concepts regarding computed tomography and the physics and mathematics at the origin of CT. As one progresses through the chapter, more detail regarding the CT technique and methodology is meet. The reader should not be alarmed if his or her particular problem or preoccupation is not mentioned here. So many different CT developments have been achieved in the last thirty-three years that it would be difficult to describe them all in a single chapter. Some practical information are also presented that can be vital to obtaining good analytical results; it is sometimes difficult to find.

I hope that this introduction to the CT technique will provide useful information to those persons who are about to get involved with CT as well as present TC users and those with simply a curiosity about the technique.

1.1. Analytical methods for image reconstruction in tomography

1.1.1. Projection and scanning of the object

The methodological basis used for describing 2D and 3D image reconstruction was presented in detail in the work of Kak and Slantay [1] and Rosenfeld and Kak [2]. Other important work such as that of Herman [3], describes the reconstruction methods such as algebraic reconstruction technique (ART). In this work, we focus on the work and analytical methods of Kak and Stanlay to explain the different steps of tomography from the measurement to the reconstruction of a single layer. The superposition of such layers will constitute the 3D image or volume representing the studied object. By means of image processing tools, parts of the reconstructed 3D volume can be extracted and analyzed separately from the rest of the data set.

An object O (x, y, z) is considered as a superposition of n layers of the same thickness along z axis, all located in planes parallel to the plane (x, y) and perpendicular to z (Fig. 1). Each layer represents a section in the object to be reconstructed. It is considered as a 2D function $f_n(x, y)$ that describes, for example, the distribution of linear attenuation coefficients as a function of the position or any other 2D function that can be measured and whose measurement signal is described by a full line. Any function maybe considered in tomography in condition to be limited and finite in a given region and equal to zero outside this region. The condition of a finite size is easily achievable for solid samples, liquid or gas contained in a box. The establishment of condition is very difficult when it will be used for the tomography of electric or magnetic fields. The generalization of such function for the tomographic measurements is closely related to the determination of the nature of the interaction of the scanning beam (comb-shaped) with the object under examination. The purpose of tomography is the reconstruction of this 2D function, representing a layer or slice of the object, from the measured projections in a unique way.

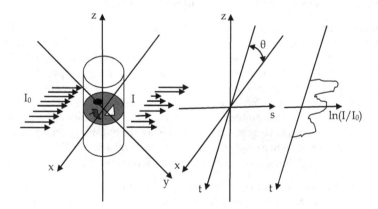

Figure 1. The geometry of a studied object scanning in the {x,y,z} coordinates system. A layer in the plane (x, y) is scanned along the angle θ and the transmitted intensity is stored in a system (t, s) of rotational coordinates.

The layer is scanned (scanned) in an angle θ varying from 0 ° to 180 ° for the transmission tomography. The intensity of the transmitted beam is recorded like a translation function of of the position parameter t (Fig. 2). The transmitted intensity is given by Lambert's law given by the following expression:

$$I(x,y) = I_0 \exp(- \int_{path} \mu(x,y)ds)$$

(1)

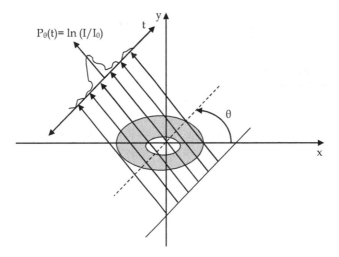

Figure 2. Scanning of a single layer in the plane (x, y). Note that z is the axis of rotation which coincides with the z-axis of the rotational coordinate's system {s, t, z}.

Where I_0 is the incident beam intensity and $\mu(x, y) = f(x,y)$ is the 2D function to rebuild. A new square and rotational coordinates system (t, s) is defined to express the detection system rotatable in comparison to the fixed object coordinates system (or vice versa, if the object is rotating and the detection system is fixed). In the transformation from the system (x, y) to the system (t, s), t is given by (see demonstration 2):

$$t = x.\cos(\theta) + y.\sin(\theta) \tag{2}$$

The expression of the ray (path) through the sample expressed in terms of t and θ and by the substitution of s is: $\delta(t-x.\cos(\theta) + x.\sin(\theta))$. The Dirac function δ ensures that only the points in obedience to the equation (2) that are related to the beam (comb-shaped) contribute to the projection P_θ (t). Such a projection can be defined by:

$$P_\theta(t)=\ln\left(\frac{I}{I_0}\right)= \int_{path} \mu(x,y)ds \tag{3}$$

Note here that the expression of P_θ (t) can be given by (see demonstration 1):

$$P_\theta(t)= \int_{path} \mu(x,y)ds$$

$$= \int\limits_{-\infty}^{+\infty} \int\limits_{-\infty}^{+\infty} \delta(x\cos\theta + y\sin\theta - t)\mu(x,y)dxdy \tag{4}$$

This last expression is just one of the different forms of the Radon transform basically used for determining a function from its integral according to certain directions (see math reminder 1). The two-dimensional Radon transform projects an object f (x, y) to get its projections P_θ (t). The projections values depend on of the integral of the object values along the line of integral according to a direction θ. In this work, we are interested in the case of a parallel beam; the extension to the case of diverging beam (fun beam) is quiet easy.

Demonstration 1

Demonstrating that the projection P_θ (t) as defined by Eq.1 can be written in one form of the Radon transform given by Eq.D.1:

$$P_\theta(t) = \int\limits_{-\infty}^{+\infty}\int\limits_{-\infty}^{+\infty} \delta(x\cos\theta + y\sin\theta - t)\mu(x,\ y)dxdy \tag{D.1}$$

For a convenience and simplicity reasons and for the fact that tomography is a process based on a rotational scanning, a new square and rotational coordinates system (t, s) is defined as presented above to take into consideration the rotating aspect of the source and the detection system around the fixed object. The coordinates of this system are given by:

$$\begin{cases} t = x.\cos\theta + y.\sin\theta \\ s = -x.\sin\theta + y\cos\theta \end{cases} \tag{D.2}$$

With:

$$dsdt = J_F dxdy \tag{D.3}$$

J_F is the Jacobian matrix determinant which is given by:

$$J_F = \begin{bmatrix} & \partial/\partial x & \partial/\partial y \\ t & \cos\theta & \sin\theta \\ s & -\sin\theta & \cos\theta \end{bmatrix} = \cos^2\theta + \sin^2\theta = 1 \tag{D.4}$$

So, in our case dsdt=dxdy, and we can write the following equation:

$$P_\theta(t) = \int\limits_{ray(\theta,t)} \mu(x,\ y)ds = \int\limits_{ray(\theta,t)} \mu(s,\ t)ds$$

$$= \int\limits_{-\infty}^{+\infty}\int\limits_{-\infty}^{+\infty} \delta(x\cos\theta + y\sin\theta - t)\mu(s,\ t)dtds \tag{D.5}$$

Note that the coordinates system transformation from (x, y) system to (t, s) system has no influence on the values of the object function μ (x, y) ((μ (x, y) = μ (s, t)).

Thus, it was demonstrated that the tomography projection can be expressed by one of the Radon's transform expression.

End of Demonstration 1.

Before continuing our mathematical development and demonstration specific to the object scanning in transmission tomography, it is useful to give some math reminder on Radon transform and its main properties.

Math Reminder 1.

The Radon transform is defined by:

$$R(p, \tau)[f(x, y)] = \int_{-\infty}^{+\infty} f(x + \tau + px)dx$$

$$= \int_{-\infty}^{+\infty}\int_{-\infty}^{+\infty} f(x, y)\delta[y - (\tau + px)]dxdy \equiv U(p, \tau), \tag{R.1}$$

where p is the slope of the projection line of and τ is its intersection with the y axis. The inverse Radon transform is given by:

$$f(x, y) = \frac{1}{2\pi}\int_{-\infty}^{+\infty} \frac{d}{dy}H[U(p, y - px)]dp, \tag{R.2}$$

where H is the Hilbert transform. The Radon transform can also be defined by the following expression:

$$R'(r, a)[f(x, y)] = \int_{-\infty}^{+\infty}\int_{-\infty}^{+\infty} f(x, y)\delta(r - x\cos a - y\sin a)dxdy, \tag{R.3}$$

Where r is the perpendicular distance of the integral line with respect to the origin and a is the angle formed by this line and the x axis.

Using the following identification:

$$F_{\omega,a}[R[f(\omega, a)]](x, y) = F_{u,v}^2[f(u, v)](x, y), \tag{R.4}$$

where F is the Fourier transform, the Radon inversion formula can be expressed by:

$$f(x, y) = c\int_0^{\pi}\int_{-\infty}^{+\infty} F_{\omega,a}[R[f(\omega, a)]]|\omega|e^{i\omega(x\cos a + y\sin a)}d\omega da \tag{R.5}$$

This last expression can be simplified as follows:

$$f(x, y) = \int_0^{\pi}\int_{-\infty}^{+\infty} R[f(r, a)]W(r, a, x, y)drda, \tag{R.6}$$

where W is a weight function given by:

$$W(r, a, x, y) = h(x\cos a + y\sin a - r) = F^{-1}[|\omega|] \tag{R.7}$$

Nievergelt (1986) determined the inversion formula as follows:

$$f(x, y) = \frac{1}{\pi}\lim_{c \to 0}\int_0^{\pi}\int_{-\infty}^{+\infty} R[f(r + x\cos a + y\sin a, a)]G_c(r)drda, \tag{R.8}$$

with:

$$G_c(r) = \begin{cases} \dfrac{1}{\pi c^2} & \text{for } |r| \le c \\ \dfrac{1}{\pi c^2}(1 - \dfrac{1}{\sqrt{1 - c^2/r^2}}) & \text{for } |r| \succ c \end{cases} \tag{R.9}$$

The Ludwig's inversion formula presents the relations between the two forms of the Radon transform of a given function: R (r, a) and R' (r, a), which are given by:

$$p = \cot\alpha, \quad \tau = \csc\alpha \tag{R.10}$$

$$r = \frac{\tau}{1 + p^2}, \quad a = \cot^{-1} p \tag{R.11}$$

The main properties of Radon transform are the followings:

1. Superposition :

$$R(p, \tau)[\, f_1(x, y) + f_2(x, y)] = U_1(p, \tau) + U_2(p, \tau); \tag{R.12}$$

2. Linearity:

$$R(p, \tau)[af(x, y)] = aU(p, \tau); \tag{R.13}$$

3. Scaling:

$$R(p, \tau)\left[\, f_1\left(\frac{x}{a}, \frac{y}{b}\right)\right] = |a|\, U\left(p\frac{a}{b}, \frac{\tau}{b}\right); \tag{R.14}$$

4. Rotation:

$$R(p, \tau)[R_\Phi f(x, y)] = \frac{1}{|\cos\Phi + p\sin\Phi|}U\left(\frac{p - \tan\Phi}{1 + p\tan\Phi}, \frac{\tau}{\cos\Phi + p\sin\Phi}\right), \tag{R.15}$$

R_Φ *is a rotation operator;*

5. Skewing:

$$R(p, \tau)[\, f(ax + by, cx + dy)] = \frac{1}{|a + bp|}U\left(\frac{c + dp}{a + bp}, \tau\frac{d - b(c + bd)}{a + bp}\right); \tag{R.16}$$

6. Integral along p:

$$I = \sqrt{1 + p^2}\,U(p, \tau); \tag{R.17}$$

7. 1D convolution expression:

$$R(p, \tau)[\, f(x, y) * g(y)] = U(p, \tau) * g(\tau); \tag{R.18}$$

8. Equivalent equation of Plancherel theorem:

$$\int_{-\infty}^{+\infty} U(p, \tau)d\tau = \int_{-\infty}^{+\infty}\int_{-\infty}^{+\infty} f(x, y)dxdy; \tag{R.19}$$

9. Equivalent equation of Parseval theorem:

$$\int_{-\infty}^{+\infty} R(p, \tau)[\, f(x, y)]^2 d\tau = \int_{-\infty}^{+\infty}\int_{-\infty}^{+\infty} f^2(x, y)dxdy; \tag{R.20}$$

End of Math Reminder 1.

At this level of mathematical development a very important question must be asked: Why the switching between different coordinates systems, presented above (Fig.1), is so important to establish the mathematical basis of image projection and reconstruction in CT in the case of analytical method such as FBP? The response to this question is satisfactory explained in the 2nd demonstration (see Demonstration 2).

Demonstration 2

Because the CT process is likely rotational in its projection phase, a rotational coordinates system (Fig.D.1) is more suitable to describe the object scanning and projection. This the first reason for using different coordinates systems.

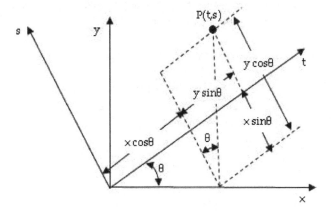

Fig. D.1. Considered coordinates system: fixed (x, y) and rotational (t, s), used to identify a projection point P of the object.

According to figure (D.1), for a given point P of the object, the transforms used to switch between a fixed coordinates system (x,y) to a rotational coordinates system (t,s) are given by:

$$\begin{cases} t = x.\cos\theta + y.\sin\theta \\ s = -x.\sin\theta + y.\cos\theta \end{cases} \qquad (D.7)$$

And the inverse transforms are given by:

$$\begin{cases} x = t.\cos\theta - s.\sin\theta \\ y = t.\sin\theta + s.\cos\theta \end{cases} \qquad (D.8)$$

According the physical principle of the projection generation in transmission CT, a projection at an angle θ can mathematically be expressed as integration of the object function (attenuation coefficient) across the line s. Thus it can be given by:

$$P_\theta(t) = \int_s f(x, y)ds = \int f(t.\cos\theta - s.\sin\theta, \ t.\sin\theta + s.\cos\theta)ds \qquad (D.9)$$

The back-projected function $f_b(x,y)$ is then given by:

$$f_b(x, y) = \int_0^\pi P_\theta(t)d\theta \qquad (D.10)$$

Where t is to be determined for each projection using equation (D.7). Thus the second reason for using rotational coordinates system is the easy representation and manipulation of Fourier transform in this system that facilitate the understanding and the implementation of the analytical reconstruction process of tomography.

End of Demonstration 2

In our case of CT, the set of all projections $P_\theta(t)$ of the object function $\mu(x, y)$ is called the Radon transform of $\mu(x, y)$. Using all these projections, a 2D image can be analytically reconstructed by exploiting a theorem called "Fourier Central Slice" (see demonstration 3). This theorem announces that the data of the 1D Fourier transform of a projection $P_\theta(t)$ is a subset of the data of 2D Fourier transform $F(u, v)$ of the object function $\mu(x, y)$:

$$FT[P_\theta(t)]=F_\theta(\omega) \subset FT[\mu(x,y)]=F(u,v), \tag{5}$$

where $u = \omega.\cos\theta$ and $v = \omega.\sin\theta$. To demonstrate this relationship, we will follow the method presented in references [1] and [2]. Assuming that the function $\mu(x, y)$ is finite and limited, so that it has a Fourier transform (u, v) given by:

$$F(u,v)=\int_{-\infty}^{+\infty}\int_{-\infty}^{+\infty} \mu(x,y).e^{-2\pi i(ux+vy)}dxdy. \tag{6}$$

To demonstrate the central slice theorem of Fourier, we consider the case of $\theta = 0$. In such a case the two coordinates systems (x, y) and (u, v) coincide and the projection $P_\theta(t)$ is simply given by:

$$P_{\theta=0}(t)=\int_{-\infty}^{+\infty} \mu(x,y)dy \tag{7}$$

The 2D Fourier transform of the object function becomes (by taking into account that $v = \omega \sin(\theta = 0) = 0$):

$$F(u,0)=\int_{-\infty}^{+\infty}\int_{-\infty}^{+\infty} \mu(x,y)e^{-2\pi i(ux)}dxdy$$
$$=\int_{-\infty}^{+\infty}\left\{\int_{-\infty}^{+\infty} \mu(x,y)dy\right\}e^{-2\pi i(ux)}dx \tag{8}$$
$$=\int_{-\infty}^{+\infty} P_{\theta=0}(x)e^{-2\pi i(ux)}dx$$

Note here that the 1D Fourier transform of a projection $P_\theta(t)$ is given by:

$$F_\theta(\omega)=\int_{-\infty}^{+\infty} P_\theta(t\equiv x)e^{-2\pi i\omega t}dt \tag{9}$$

Thus, we have demonstrate that $F_{\theta=0}(\omega)$ is just a subset of $F(u, 0)$ for a particular case of $\theta=0$. Because the orientation of the object in the coordinates system (u, v) is arbitrary with respect to the coordinates system (x, y), this particular case can be extended to all angles θ by keeping in mind that the Fourier transform is conserved by rotation which is necessary process in transmission tomography in the real space. Indeed, the 1D Fourier transform of $P_\theta(t)$ produces a values which are a part (the same) of the values produced by the 2D Fourier transform of $\mu\,(x, y)$.

A dense set of values in the Fourier space can be approximated and simplified by considering square coordinates that will easily back-transformed into real space. This can, sometimes, lead to a confused reconstruction which is the consequence of an unachieved adjustment of the high frequencies in the Fourier space (Fig. 3).

Figure 3. Fourier transform of a projection $P_\theta(t)$ (left) and the interpolated data in the square coordinates system (right). The interpolation of high frequencies (high values of (u, v)) is inaccurate.

Demonstration 3.

Here, we demonstrate the Fourier Central Slice Theorem announcing that the Fourier transform of a projection $P_\theta(t)$ is a subset of a of two-dimensional Fourier transform of the $\mu\,(x, y)$ object function (Fig. D.2).

The Fourier transform of a projection $P_\theta(t)$ may be given by the following expression:

$$P_\theta(\omega) = \int_{-\infty}^{+\infty} P_\theta(t)e^{-2\pi i \omega t}\,dt \qquad (D.11)$$

The projection is given as mentioned before by:

$$P_\theta(t) = \int_{-\infty}^{+\infty}\int_{-\infty}^{+\infty} \delta(x\cos\theta + y\sin\theta - t)\mu(x, y)dxdy \qquad (D.12)$$

The last expression of $P_\theta(t)$ can be whiten in the polar coordinates system (r, φ) be the following expression:

$$P_\theta(t) = \int_0^{2\pi}\int_0^\infty \delta(r\cos\varphi\cos\theta + r\sin\varphi\sin\theta - t)\mu(r, \varphi)|r|drd\varphi$$
$$= \int_0^{2\pi}\int_0^\infty \delta(r\cos(\varphi-\theta) - t)\mu(r, \varphi)|r|drd\varphi \qquad (D.13)$$

If we take the 1D Fourier transform of last expression of $P_\theta(t)$, we get the following expression:

$$P_\theta(\omega) = TF(P_\theta(t)) = \int_{-\infty}^{+\infty} P_\theta(t)e^{-2\pi i\omega t}\,dt$$

$$= \int_{-\infty}^{+\infty}\int_0^{2\pi}\int_0^\infty \delta(r\cos(\varphi-\theta)-t)\mu(r,\,\varphi)e^{-2\pi i\omega t}\,|r|\,dr\,d\varphi\,dt \qquad (D.14)$$

$$= \int_0^{2\pi}\int_0^\infty \mu(r,\,\varphi)e^{-2\pi i\omega r\cos(\varphi-\theta)}\,|r|\,dr\,d\varphi$$

Similarly, we proceed to the calculation of the 2D Fourier transform of the object function $\mu(x, y)$.

$$F(u,\,v) = \int_{-\infty}^{+\infty}\int_{-\infty}^{+\infty}\mu(x,\,y)e^{-2\pi i(ux+yv)}\,dx\,dy \qquad (D.15)$$

If we transform this last expression to polar coordinates in both spatial and frequency domains using the following well-known transformation equations:

$$\begin{cases}x = r\cos\varphi \\ y = r\sin\varphi\end{cases} and \begin{cases}u = \omega\cos\psi \\ v = \omega\sin\psi\end{cases}, \qquad (D.16)$$

we obtain the following expression:

$$F(u,\,v) = F(\omega\cos\psi,\,\omega\sin\psi)$$

$$= \int_0^{2\pi}\int_0^\infty \mu(r,\,\varphi)e^{-2\pi i\omega r(\cos\varphi\cos\psi+\sin\varphi\sin\psi)}\,|r|\,dr\,d\varphi \qquad (D.17)$$

$$= \int_0^{2\pi}\int_0^\infty \mu(r,\,\varphi)e^{-2\pi i\omega r\cos(\varphi-\psi)}\,|r|\,dr\,d\varphi$$

The comparison between the equations (D.13) and (D.16), allow us to write the following equivalence:

$$TF_{1D}[P_\theta(t)] = P_\theta(\omega) = TF_{2D}[\mu(r,\,\varphi)] = F(\omega\cos\psi,\,\omega\sin\psi)_{\theta=\psi} \qquad (D.18)$$

Thus, it was demonstrated that the 1D Fourier transform of a projection is a subset (central slice) of the 2D Fourier transform of the object function. Therefore, a simple back-transformation (inverse transform) of the Fourier transform may allow the reconstruction of a 2D layer (slice) of the object.

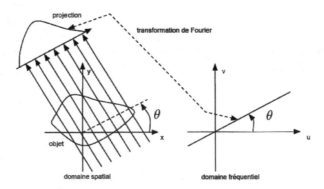

Fig. D.2. Illustration of Central Slice Theorem of Fourier.

Image reconstruction using the central slice theorem is theoretically possible for an infinite number of projections. For real data case, we have only a finite number of projections. In this case, the Fourier transform function F(u,v) is known only on number points along radial lines. In practice, the number of samples (points) taken is the same for each projection direction. As well, in the Fourier domain, the sampling is constant regardless of the direction of projection.

Thus, the digitalization and sampling is a very important process in practical tomography. To be able to reconstruct the object function, these samples (points) must be interpolated from a polar to a Cartesian coordinates (Fig. D.3). Generally, this interpolation is done by taking the nearest neighbour value or by a linear interpolation between known points. The density of points (frequencies) in the polar reference becomes smaller when we move away from low frequencies (i.e., the origin). So the interpolation error is larger at high frequencies than at low frequencies, and this causes the degradation of the image details.

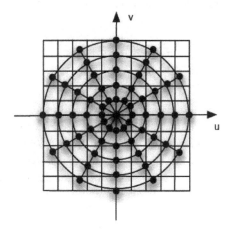

Fig. D.3. Transition from a polar grid to a Cartesian square grid.

End of Demonstration 3

How many projections are needed for good reconstruction? This question has been answered by the Nyquist-Shannon theorem. This theorem announced that a unique reconstruction of an object sampled in space is obtained if the object was sampled with a frequency greater than twice the highest frequency of the object details. In the parallel scan mode, for a sampling for S points per projection line, a number of P projections (angles θ) is necessary to accomplish the Shannon theorem in tomography. If D is the diameter of the object to be scanned, and Ax is the difference between two points of scanning, then the number of points (sampling) in each projection line to be scanned is given by:

$$S = \frac{D}{\Delta x} \tag{10}$$

For a scanning of the object over 360 °[1], each point is scanned again after a path equal to πD and this for each point situated on the surface of a circular object of a diameter D (the highest frequency). For this case, the number of projections must be equal to:

---

1 Generally if the object is homogenous and symmetric, a scanning over 180° is sufficient.

$$P = \frac{\pi D}{\Delta y} \tag{11}$$

According to Nyquist-Shannon's theorem it is required for a good reconstruction that $\Delta y \geq 2\Delta x$. If we put $\Delta y \geq 2\Delta x$, we obtain a relationship between the number of scanned points S in one projection line (sampling of the object) and the number of necessary projections P (see demonstration 4):

$$P \geq \frac{\pi}{2} S \tag{12}$$

Indeed, S and P are the most important parameters that determine the quality of the reconstruction. If P violates the last condition (Eq.11) when for example a number less than necessary P projections is recorded, a new reduced diameter (D = D *) which satisfies the Shannon condition must be calculated. This reduced dimater determines the reduced volume (area) of the object that can be well reconstructed for this reduced number of projections although the whole object of real diameter D is scanned.

Demonstration 4.

In computed tomography, it is recommended that the number of projections (P) must be in the same order as the number of rows of pixels in a single projection (S) [3]. The condition on P and s given by Eq.12 which is established when considering the Nyquist-Shannon sampling theorem can be proven by the following manner. Let considering P projections over 180° and S rows of pixels, the angular increment between two successive projections $\Delta\theta$ in the Fourier space is given by [4]:

$$\Delta\theta = \frac{\pi}{P} \tag{D.18}$$

For a distance Δx between two adjacent rows, the highest spatial frequency measured (ω_{max}) in a projection is given according to the Nyquist-Shannon theorem by [5]:

$$\omega_{max} = \frac{1}{2\Delta x} \tag{D.19}$$

The scanned object representation in the frequency domain (2D Fourier Transform) correspond to disk's radius that contains the measured values (Fig. D. 4). The distance Δf between two consecutive values of the furthest circle from the origin is given by:

$$\Delta f = \omega_{max}\Delta\theta = \frac{1}{2\Delta x}\frac{\pi}{P} \tag{D.20}$$

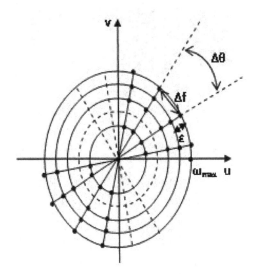

Fig. D.4. Density of the measured values in the frequency domain.

For the S values of each projection in spatial domain correspond S measured values in the frequency space (measured for each line). Therefore, the distance ε between two consecutive values measured on a radial line in the frequency domain (Fig. D.4) is given by:

$$\varepsilon = \frac{2\omega_{max}}{S} = \frac{1}{DS}$$

(D.21)

A sufficient condition to obtain a good reconstruction is to ensure that the worst azimuth resolution (s) in the frequency domain (Fig. D.4) is in the same order of radial resolution (ε). This condition can be expressed as follows:

$$\frac{1}{2D}\frac{\pi}{P} \approx \frac{1}{DS} \Rightarrow P \approx S\frac{\pi}{2}$$

(D.22)

Thus, the ratio between the number of projections (P) and the number of rows (S) must be in the order of π / 2.

An insufficient number of projections may produce a 2D or 3D reconstructed image which present many undesirable artefacts. In practice, most of the tomography detectors cannot measure below the nominal Nyquist resolution determined by their dimensions or pixel size. The exploration beam is not perfectly parallel, too; this is another source of errors on the measured data obtained especially when using a reconstruction method which assumes that the beam is parallel.

In practice, insufficient number of projection generates artefacts such as those shown in figure (D.5) [3]. The projections of figure (D.5) have a dimension of 64x64 pixels. So the number of rows S is 64. As it has already been demonstrated, a number P of projections around a value of 100 (P ≈ Sπ / 2) is sufficient to produce an acceptable reconstructed image. For the case of this example, it is clear that the reconstruction image obtained from 64 projections is the closest one to the original image. The 2D reconstruction is performed using the Filtered Back-Projection method (FBP) that will be described in the next section.

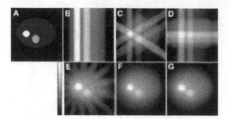

Fig. D.5. Results of 2D reconstruction of an object by FBP method of tomography and induced artifacts for different number of projections: (A) original image, (B) 1 projection, (C) 3 projections, (D) 4 projections, (E) 16 projections, (F) 32 projections and (G) 64 projections [7] .

The number of projections is not the only parameter that affects the reconstructed image. The sampling of projections S (number of rows) and the dimensions of the grid of reconstruction will also affect the reconstructed image dramatically if they are not well optimized

End of Demonstration 4

1.1.2. 2D and 3D image reconstruction

As already mentioned, the inverse transformation (simple inversion) of the Fourier transform $P_\theta(\omega)$ can be directly used to produce the 2D reconstructed layer of μ (x, y). However, a more efficient way and more elegant has been developed called " Filtered Back-Projection (FBP)" making the reconstruction process less expensive and less complicated when compared to the direct calculation of the inverse Fourier transform [1,2]. The basic idea of this method is derived from the tomography principle and the scanning mode: the object (set of layers) is scanned projection by projection from 0 ° to 180 ° which means that $P_\theta(t) = P_{\theta+180}(-t)$ and suggesting the use of a polar coordinates rather than Cartesian square ones. As it was demonstrated the digitization and sampling details by the right selection of P and S are very important in Computed tomography. If the projections are recorded in polar coordinates, the projections Fourier transforms will be discrete values of a polar function. Thus, it seems appropriate to write the 2D Fourier transform of the object function μ (x, y) in polar coordinates. The inverse Fourier transform μ (x, y) of F(u, v) written in Cartesian coordinates is given by:

$$\mu(x,y)=\int_{-\infty}^{+\infty}\int_{-\infty}^{+\infty}F(u,v).e^{i2\pi(ux+vy)}\ dudv \tag{13}$$

The same inverse transform in polar coordinates is given by:

$$\mu(x,y)=\int_{0}^{2\pi}\int_{-\infty}^{+\infty}F(\omega,\theta).e^{i2\pi\omega(x\cos\theta+y\sin\theta)}\ \omega d\omega d\theta \tag{14}$$

The substitution of x cos (θ) + y sin (θ) by t and the application of Fourier Central Slice theorem, allow us to get the following expression:

$$\mu(x,y)=\int_0^\pi \int_{-\infty}^{+\infty}[F(\omega,\theta).e^{+2\pi i\omega t}\mid\omega\mid d\omega]d\theta$$

$$=\int_0^\pi \left[\int_{-\infty}^{+\infty} F(\omega,\theta).e^{+2\pi i\omega t}\mid\omega\mid d\omega\right]d\theta \tag{15}$$

The integral in brackets can be regarded as the inverse Fourier transform $P_\theta(t)$ of $P_\theta(\omega)$. However, we note that it is multiplied by the function $\mid\omega\mid$ which plays the role of a special ramp filter function in the frequency domain. So, we define the filtered projection by:

$$Q_\theta(t)=\int_{-\infty}^{+\infty} P_\theta(\omega)\mid\omega\mid e^{+2\pi i\omega t}d\omega \tag{16}$$

Thus, the object function will be simply given by:

$$\mu(x,y)=\int_0^\pi \left[Q_\theta(t)\right]d\theta=\int_0^\pi \left[Q_\theta(x\cos\theta+y\sin\theta)\right]d\theta \tag{17}$$

Knowing previously that a product (multiplication) in Fourier space (frequency domain) corresponds to a convolution of inverse Fourier transforms in real space (spatial domain), the following relation can be written:

$$FT^{-1}[P_\theta(\omega)\times\mid\omega\mid]=P_\theta(t)\otimes FT^{-1}[\mid\omega\mid] \tag{18}$$

The convolution operator is denoted by the symbol $\otimes$. The function $\mid\omega\mid$ is not a square-integrable function, thus it has no inverse Fourier transform. However, the inverse transform $FT^{-1}[\mid\omega\mid]$ can be approximated by different filter response functions (convolution with gains, filter functions). Considering these last remarks and regarding the result of Eq.17, The reconstruction process of $\mu(x, y)$ can be performed from the projections $P_\theta(t)$ obtained as follows:

"In FBP reconstruction process, each projection $P_\theta(t)$ is first convoluted with a specific and suitable filtering function (eg Shepp-Logan, $\omega/2\pi$ sinc(ω)) and the measured values are recorded in (x, y) plans as illustrated in Fig. 4. To control the measurement process of tomography, the projections are arranged in a sinograms[2] recorded as shown in Fig. 5. On the sinogram, each vertical line is a projection at an angle θ and represents the variation in the gray level as a function of the pixel's or detector's position. On Fig. 6 are shown 4 different reconstructions. The projections $P_\theta(t)$ are converted to grayscale, convoluted with a filter with a specific gain and back-projected onto the entire plan (x, y). The addition of all projections result in the reconstruction of the 2D layer desired. More the number of projections is high more the reconstruction plans is well covered in terms of data and less the star-shaped artefacts are present on the reconstructed image. With such reconstructions, 3D images can be obtained by stacking all the 2D layers and a reconstructed 3D volume data (details of the object) can be extracted from the stack obtained".

---

2 the number of sinograms is equal to the sampling number of S

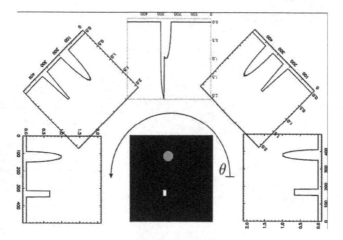

Figure 4. Scanning a 2D object (details: circle and square) and the corresponding projections [7].

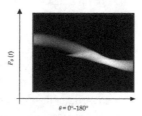

Figure 5. Example of projection Sinogram of the object of Fig. 4: the y-axis shows the projections (gray level variation as a function of the pixel's or detector's position) and on the x-axis, these one pixel width projections are arranged from the first projection (θ=0) to the last one (θ=180°). Each sinogram will be used for the reconstruction of one pixel layer of the object.

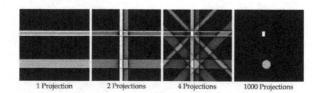

Figure 6. Reconstructions obtained for 4 different numbers of projections to illustrate the appearance of star-shaped artefacts when the number of projection is not sufficient.

1.1.3. Discretization of the analytical methods

In the last sections, many times we have considered the ideal continuous case to model the projection and the reconstruction processes of tomography and to explain the principle of the analytical reconstruction methods such as FBP. Indeed, we have worked in an infinite continuous domain (R^2), and tried to reconstruct a continuous function f from continuous projections P_θ defined for all angle θ in the interval $[0, \pi\,[$. These conditions are obviously not possible in practice. Acquisition systems allow just obtaining a number of projections P_θ for a finite number of angles, denoted θ_k. The limited number of detectors makes these projections sampled and known only at discrete points u_k. It would be unrealistic with such data to try to reconstruct a continuous function f in R^2. Therefore, the object function f(x,y) will be reconstructed on a discrete grid for a finite number of points ($f(x_i,y_j)$). This is also the limit imposed by the used numerical reconstruction algorithms. The reconstruction problem then will be approached as follows: for given a set of projection measurements:

$$\left\{P_{\theta k}(u_l), 0 \le 1 \le S, 0 \le k \le P\right\},\tag{19}$$

the reconstruction problem is reduced to finding f at any point of a finite discrete grid:

$$\left\{f(x_i,y_j), 0 \le i \le S, 0 \le j \le S\right\}\tag{20}$$

with:

$$\begin{aligned}\theta_k &= k\Delta\theta, \quad \Delta\theta = \pi/P, \\ x_i &= i\Delta x, \quad y_j = j\Delta y,\end{aligned} \quad u_l = ld,\tag{21}$$

where $\Delta\theta$ is the sampling step of the rotation angles, P is the number of angles (projections), d is the sampling step on each projection line (ray), Δx and Δy are the sampling step on x and y in the reconstruction plan.

The discrete reconstruction methods according to this definition can be divided into two categories:

1. The first class method consists in the definition of discrete operators and functions equivalent to those defined in the continuous case (Radon transform, back-projection, Fourier transform, etc.) and all the inversion formulas and theorems defined for the analytical methods already presented.

2. The second category is based on a completely different approach: in these methods the projection equation (p=Rf) is directly discredited and a linear equations system is built. The resolution of this system is possible only by iterative methods called algebraic methods. The algebraic method will be presented in the next section.

2. Algebraic methods of image reconstruction in tomography

The algebraic methods are especially iterative methods. These methods are less used when compared to the easy and well known filtered back-projection method. In this method the reconstruction problem is see differently and no longer refers to the Radon transform. The image of the object consists of a number k of pixels whose values f_k are unknown. Similarly, the projections are discrete and formed by a number of l dexels (depth pixels) whose values "p_l" are known since they correspond to measurements in each line of projection. Reconstruction of the object image by the iterative method is based on the following hypothesis: each detected values in a dexel is a linear combination of pixel's values to be reconstructed [6]. The reconstruction problem is formulated by a discrete expression of matrix (p = R.f) describing the projection process. The set of values of the projection lines (dexels) is arranged in projection vector p. All pixels of the image to be reconstructed are also grouped in a image vector f. The coefficients which characterize the contribution of each pixel in each line of projection is determined and stored in a matrix R. The projection of an object is modeled in the case of an original image of n x n pixels, a number n of projection directions and a number n of dexels in each projection line, by the following equation [6]:

$$p = R.f \tag{22}$$

$$
\begin{bmatrix} p_1 \\ p_2 \\ \vdots \\ \vdots \\ p_n \end{bmatrix}
=
\begin{bmatrix} r_{11} & \cdots & \cdots & \cdots & r_{14} \\ \vdots & \ddots & & & \\ \vdots & & \ddots & & \\ \vdots & & & \ddots & \\ r_{41} & \cdots & \cdots & \cdots & r_{44} \end{bmatrix}
\cdot
\begin{bmatrix} f_1 \\ f_2 \\ \vdots \\ \vdots \\ f_n \end{bmatrix}
\tag{23}
$$

This last equation expresses the fact that what is detected (p) is the result of values (f) of the image to be reconstructed, subject to a projection operation represented by the projection operator R [6]). Through this modeling of the projection process, one looks in practice to find f according to p by solving the inverse problem $f = R^{-1}.p$. Because of the size of this equations system, the resolution cannot be performed except by successive iterations [7]. Fig. 7 shows an example of projection and back-projection. In the algebraic reconstruction method based on iterative process, the back-projection is modelled by a back-projection operator R^t who is none other than the transposed matrix of R. Thus, the reconstruction problem is limited in solving the inverse problem $f = R^t.p$.

The resolution of the inverse problem by iterative methods consists in finding a solution f minimizing the distance d between p and R.f where p and R are known. Here, it is question to start from an arbitrary estimate of the image solution and to proceed schematically to the correction the first estimate basing on a principle of trial and error. Each next estimate is projected and the result obtained is compared to the measured projection. The returned error is used to improve the next estimate. This method leads to build gradually low-to-high frequencies of the image solution. The results of the first iterations are smooth because of the predominance of low frequencies (internal structure) of the object. Subsequently, more iteration are applied more high frequencies (overall shape and background noise) are represented. The image produced by iterations approximate gradually the image solution (the algorithm converges) [6]. However, we show that when using such iterative methods and after a number of iterations, the process begins to diverge (under the influence of noise) and the image moves away from the true solution. To overcome this inconvenient, we impose a constraint to the reconstruction process to interrupt the iteration after a certain number of iterations. This is equivalent to use a low pass filter as in the case of filtered back-projection method.

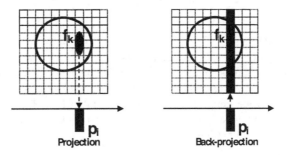

Figure 7. Illustration of projection and back-projection processes.

2.1. Projection modeling

The projection process is modeled by considering the coefficients of the projection matrix R that generate the acquisition data. Some special geometric and physical considerations are necessary for this modeling. In the iterative reconstruction method the modeling concerns the following points:

1. **Modeling of the pixel (detector) intensity distribution**: it is necessary to specify the real conditions of image pixels projection [6]. It is based on the evaluation of the contribution of each pixel in the corresponding projection line (ray). Knowing that the most accurate model (perfect) and the more complicated to be applied consists in considering square pixel (uniform); simpler models are possible. Among these simple models, there is the model called Dirac model in which all the pixel intensity is concentrated in the centre of the pixel. Thus, the whole intensity of the pixel contributes to the projection line (ray) if and only if it passes through the Dexel. There is also another model called the model of concave disc which is considered as a compromise between the two previous models. In this model, the intensity of the pixel is geometrically limited to a disk included in the pixel and distributed so that its projection is rectangular regardless of the direction of projection [6].

2. **Geometric modeling of the projection operator**: for the determination of the coefficients of the projection matrix R, we must consider the number of projections and their angular distributions and the projection beam geometry which can be parallel or fan. If, for example, the intensity model distribution is that of Dirac, a given pixel with an index k crossed by a ray of an index p_1 generates a projection coefficient r_k equal to 1; if this is not the case, zero value is assigned.

3. **Physical modeling of the projection operator**: this model is based on the distance between the position of the object pixel to the detector. A pixel located away from the detector will see its contribution to the projection ray reduced compared to a nearest one. With this modeling aspect the beam attenuation will be included in the resulting reconstructed image.

2.2. Main iterative algorithms

There are many algorithms that have been developed that used the iterative method for tomographic reconstruction (see practical example 1). Among these algorithms, the main ones are: the Algebraic Reconstruction Technique (ART), the Simultaneous Iterative Reconstruction Technique (SIRT), and the Iterative Least Squared Technique (ILST). Currently, the most commonly used algorithms are: the Expectation Maximization algorithm (EM) and the Conjugate Gradient algorithm (CG).

1. **EM algorithm**

This algorithm was developed and proposed by Lange and Carson. The formula of this algorithm is the following [6]:

$$f^{n+1} = f^n R^t \frac{p}{R.f^n} \tag{24}$$

Where n is the number of the actual iteration. This algorithm is characterized by the fact that it keeps the number of iterations for each projection. Moreover its multiplicative form gives a positivity constraint although it implies a slow convergence.

2. **Conjugate gradient algorithm (CG)**

It is an algorithm which is used because it converges very quickly. It is based on a classical descent method. Its formula of iterative updating can be, roughly, given by the following expression [7]:

$$f^{n+1} = f^n + \alpha^n d^n \tag{25}$$

We can verify that the correction is not multiplicative as for EM algorithm, but it is additive. This formula is characterized by a descent direction d (Eq. 26) and a descent speed α (Eq.27) that are recalculated in a conjugated manner for each iteration and this to optimize the speed of convergence [6].

$$\alpha^n = \frac{\left\|R^t p - R^t R f^n\right\|^2}{(R^t p - R^t R f^n)^t R(Rt\ p - Rt\ R\ fn)} \tag{26}$$

$$\begin{aligned} &iteration.1: & d^1 &= R^t p - R^t R f^0 \\ &iteration.2: & d^n &= (R^t p - R^t R f^n) + b^n d^{n-1} \end{aligned} \tag{27}$$

Through this process, the error between the measured projections and those calculated is minimized progressively. This error evaluation formula is given by:

$$e = \left\|p - Rf\right\|^2 \tag{28}$$

3.   ART algorithm

The formula of the iterative updating of this algorithm is given by the following expression:

$$f_i^{(n+1)} = f_i^{(n)} + R_{ji}\frac{p_j - R_j f^{(n)}}{\left\|R_j\right\|^2} \tag{29}$$

The correction in this algorithm can be additive ($e=p_k-p_k^n$) or multiplicative ($e= p_k/p_k^n$).

4.   SIRT algorithm

The formula of the iterative updating of this algorithm is given by the following expression:

$$f_i^{(n+1)} = f_i^{(n)} + \frac{\sum_j p_j}{\sum_j \sum_i R_{ji}} - \frac{\sum_j R_j f^{(n)}}{\sum_j \left\|R_J\right\|^2} \tag{30}$$

Practical example 1

To illustrate the difference between the iterative algorithms, let consider the example of figure (E.1). In this example we consider an object (image) composed by 2 pixels which are projected as shown on this figure.

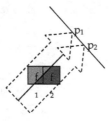

Fig. E.1. Object (2 pixels) and projection model

For each projected line, the projector matrix is modeled by considering a value proportional the area of the pixel covered by the ray (projection line).For example for p_1 the projection ray cover approximately ¾ of the area of the first pixel (f_1) and ¼ of the area of the second pixel (p_2). Therefore, the projection operator (matrix) can is given by:

$$R = \begin{bmatrix} 3/4 & 1/4 \\ 1/4 & 3/4 \end{bmatrix} \tag{E.1}$$

Supposing that the object function in known ($f_1=2, f_2=3$). In this case the projection values are equal to:

$$p_1 = \tfrac{3}{4}f_1 + \tfrac{1}{4}f_2 = \frac{9}{4}(1)$$
$$p_2 = \tfrac{1}{4}f_1 + \tfrac{3}{4}f_2 = \frac{11}{4}(2) \tag{E.2}$$

We try know to compute the object function data (f_1, f_2) using some practical iterative algorithms and to compare the values obtained to real ones ($f_1=2, f_2=3$). For all of the following algorithms, the initial iteration conditions are: $f_1^0 = f_2^0 = 0 ="/> p_1^0 = p_2^0 = 0$.

1. Additive ART algorithm with Kackzmarz method

In this method, the iteration updating is given by:

$$f_i^{n+1} = f_i^n + (p_k - p_k{}^n)^{R_{ki}} \Big/ {\textstyle\sum_j R_{kj}{}^2} \tag{E.3}$$

During the iteration process just one ray equation (p_1 or p_2) is used per iteration alternatively. Results of the iteration and p_1, p_2 updating are shown on the following matrix.

Iteration(n)	1	2	3	4	5	6	7	8	9	10	11	12	13
f_1^n	2.70	3.26	2.25	2.45	2.20	2.25	2.06	2.10	2.02	2.02	2.02	2.02	2.00
f_2^n	0.90	2.58	2.24	2.84	2.75	2.90	2.84	2.96	2.93	2.98	2.97	2.99	2.98
p_1^n	/	3.09	/	2.54	/	2.41	/	2.31	/	2.27	/	2.26	/
p_1^n	1.35	/	2.24	/	2.61	/	2.64	/	2.70	/	2.73	/	2.73

$$\text{(E.4)}$$

We can easily verify that this algorithm gives a solution which approximate well the real one (2,3) after 12 iterations (2.02, 2.99)

2. SIRT algorithm with Jacobi method

In this method, the iteration updating is given by:

$$f_i^{n+1} = f_i^n + (p_k - p_k{}^n) / R_{ii} \tag{E.5}$$

In this method f_1^n is calculated from the equation of $P_1(Eq.E.3(1))$ and f_1^n from the equation of $P_1(Eq.E.3(2))$ for each iteration by taking into consideration p_1^n and p_2^n updating. The results obtained are shown n the following matrix:

Iteration(n)	1	2	3	4	5	6	7	8	9	
f_1^n	3	1.78	2.99	2.73	2.11	2.08	2.01	2.01	2.00	
f_2^n	3.66	2.66	0.81	2.68	2.75	2.96	2.97	2.99	2.99	(E.6)
p_1^n	3.16	2	2.44	2.71	2.27	2.30	2.25	2.25	/	
p_2^n	4.49	2.44	1.35	2.69	2.59	2.74	2.73	2.74	/	

With this algorithm and method the best solution is obtained after 9 iterations.

3. SIRT algorithm with Gauss-Seidel method

This method is similar to the Jacobi method and use the same updating function (E.5) but just for the first iteration which allows the determination of f_1^1 value. After that f_2^1 is estimated from p_2 equation considering the obtained value f_1^1 which itself will be used to determinate f_1^2 from p_1 equation and so on (by matching alternatively between p_1 and p_2) until obtaining the best results. The results obtained are shown on the following matrix:

Iteration(n)	1	2	3	4	
f_1^n	3	1.18	1.90	1.98	(E.7)
f_2^n		2.66	3.27	3.03	3.00

We remark that this algorithm and method gives good results just after 4 iterations; so it is the very fast one in term of convergence.

End of practical example 1

2.3. Iteration stopping rules

The iterative process for image reconstruction must be stopped at the end of the convergence phase before it starts diverging. This can be done using some regularization methods. Indeed, well selected regularization parameter can controls the amount of stabilization imposed on the solution. In iterative methods one can use the stopping index as regularization parameter. When an iterative method is employed for image reconstruction, the user can also study on-line adequate visualizations of the iterates as soon as they are computed, and simply halt the iteration when the approximations reach the desired quality. This may actually be the most appropriate stopping rule in many practical applications, but it requires a good intuitive imagination of what to expect. If this is not the case, a computer can give us some aid to determine the optimal approximation. The stopping rules are divided into two categories: rules which are based on knowledge of the norm of the errors, and rules which do not require such information. If the error norm is known within reasonable accuracy, the perhaps most well known stopping rule is the discrepancy principle Morozov [8]. Examples of the second category of methods are the L-curve criterion [9], and the generalized cross-validation criterion [10].

3. Conclusions

After a substantial effort, major breakthroughs have been achieved in the last fourteen years in the mathematical modeling of CT. The aim of this chapter is to survey this progress and to describe the relevant models, mathematical problems and reconstruction procedures used in CT. We give a summary on the mathematics of computed tomography. We start with a short introduction to integral computed tomography. We then go over to projection and inversion algorithms. We give a detailed analysis of the filtered back-projection algorithm in the light of the sampling theorem. We also describe the convergence properties of iterative algorithms. We shortly mention Fourier based algorithms used in CT.

Good reconstructions from the interlaced lattice can also be obtained by using the direct algebraic reconstruction algorithm, or by increasing the amount of data through the interpolation according to the sampling theorem. As we have seen, the interpolation step can introduce significant errors in certain cases. It has also been shown that the interpolation can be avoided by choosing the points x where the reconstruction is computed on a polar grid rather than on a square Cartesian grid, and interchanging the order of the two summations. This algorithm should work well for the interlaced lattice and is particularly beneficial in case of the fan-beam sampling geometry, since the method also avoids the homogeneous approximation, whose influence on the reconstruction is difficult to estimate.

The cone and fan beam scanning are the standard scanning modes in present day's clinical practice. The methods described above assumed that the geometry of the acquisition was a parallel geometry, like that of first generation systems. In the case a conical geometry (or fan), three methods are possible:

1. The first is to ignore the discrepancy. The error induced by this approximation is considered negligible if the beam angle is low (typically below 15 degrees). This method is applicable to systems of second generation, but more to the following where the beam must cover the whole section of the patient to remove the translation.

2. The second method is to rearrange the data in parallel projections, mainly by interpolation.

3. The last method is to reformulate the problem completely. It becomes apparent that the projection theorem cannot be generalized, which does not have direct inversion method. The projection theorem can be adapted to different geometries, and then the algorithm is the same as for a parallel geometry, with the same calculations. Finally, the filtered back projection formulas can be corrected and result in a slightly different algorithms. The algebraic methods have some fewer additional problems, but the flexibility in the choice of basic functions and thus the coefficients of the matrix R, allows these methods to be adapted to different geometries.

Finally, I want just to add that the purpose of this chapter is to give an introduction to the studied topic and treat some related aspects in more detail. The reader interested in a broader overview of the field, its relation to various branches of pure and applied mathematics, and its development over the years may wish to consult the appropriate bibliography.

Author details

Faycal Kharfi

Department of Physics, Faculty of Science, University of Ferhat Abbas-Sétif, Algeria

References

[1] Kak, A. C, & Slaney, M. Principles of Computerized Tomographic Imaging. N.Y: IEEE Ed; (1999).

[2] Rosenfeld, A, & Kak, A. C. Digital Picture Processing, Computer Science and Applied Mathematics. Academic Press Inc; (1982).

[3] Herman, G. T. Fundamentals of Computerized Tomography: Image Reconstruction from Projections. Springer; (2009).

[4] Schillinger, B. Neutron Tomography, PSI summer school on neutron scattering. Switzerland ; (2000).

[5] Ouahabi, A. Fondements théoriques du traitement de signal. Alger : Connaissance du Monde ; (1993).

[6] Darcourt, J. Méthodes itératives de reconstruction, Revue de l'ACOMEN, N°2 ; (1998). , 4

[7] Buvat, I. Reconstruction Tomographique : www.guillement.org/ireneaccessed 3 July (2008).

[8] Morozov, V. A. On the solution of functional equations by the method of regularization. Moscow: Soviet Math. Dokl. 7; (1966). , 414-417.

[9] Hansen, P. C. Rank-Deficient and Discrete Ill-Posed Problems. SIAM; (1998).

[10] Golub, G. H. M, & Heath, G. Wahba, Generalized cross-validation asa method for choosing a good ridge parameter. Technometrics, 21 ((1979). (2), 215-223.

Chapter 5

Transparent 2d/3d Half Bird's-Eye View of Ground Penetrating Radar Data Set in Archaeology and Cultural Heritage

Selma Kadioglu

Additional information is available at the end of the chapter

1. Introduction

This chapter focuses a new visualization approach to the monitoring of internal micro-discontinuities such as cracks, micro-fractures and cavities, and archaeological remains with foundational infrastructure. The method uses a hybrid interactive two-dimensional / three-dimensional (2D/3D) transparent visualization of ground penetrating radar (GPR) data set gathered from sites of archaeological and cultural heritage. The data visualization is based on methodological formulation of amplitude–colour scale function for 2D radargram visualization to indicate micro-discontinuities, infrastructures and buried archaeological remains. The transparent 3D imaging combines to half bird's-eye view was constructed from a processed parallel-aligned 2D GPR profile data set by using an opaque approximation instead of linear opacity. The amplitude–colour scale is balanced by the amplitude range of the buried remains within a proposed depth range, and appointed an opaque coefficient in order to differentiate buried remains from others. Interactive visualizations are conducted of transparent 3D half bird's-eye view of GPR data volumes.

Archaeological and cultural heritage represent cultural identity and a source of creativity for present and future generations. Conservation of art treasures is a serious problem, since all objects change or deteriorate over time, mainly due to natural forces of decay. Requirements for maintaining the existing condition of buildings, monuments or statues of historical interest differ from the requirements of an initial treatment. Many of the problems associated with treatment involve the lack of prior, baseline information. Identification of the causes of degradation and understanding of the cause/effect relationships are crucial for the conservation of archaeological and cultural heritage. The treatment of such items generally involves coating or vapour barriers, and compatibility with substrate. However, cultural heritage requires maintenance not only for its walls, but also its infrastructures and security of foundations. Therefore, safety and maintenance management of such sites must include imaging of infrastructures and potential discontinuities. The same importance should be given to archaeological sites, especially in urban areas. Therefore, there is also a need for improved non-invasive methods of visualization in evaluating the progress of the buried infrastructures of archaeological heritage.

Ground penetrating radar (GPR), which is also called surface penetrating radar, is a time-dependent, high frequency electromagnetic geophysical technique that can provide a 3D pseudo image of the subsurface, including the fourth dimension of colour, and can also provide accurate depth estimates for many common subsurface objects [1, 2]. GPR uses the scattered wave field of high frequency electromagnetic (EM) waves. The EM waves travel at a specific velocity that is determined primarily by the permittivity of the material. The principles of GPR have been explained extensively in the literature [3], especially for fault and fracture imaging [4-9]; in assisting contaminated sites by locating buried features of interest such as underground storage tanks, pipes [10-11], unexploded ordnance (UXO) and clutter [12-14]; and in the mapping of shallow stratigraphy and discontinuities [3, 15-20].

Ground penetrating radar (GPR) provides more detailed results than other geophysical methods, because it can image the position and the depth of targets within very complex and restricted areas. The method is non-destructive and can be applied on a surface, a wall, or a monument [21-23]. The method can also be used in urban areas or in archaeological structures and, depending on the antennas and the particular situation, can achieve a resolution of the order of several centimetres 24-26]. Therefore, it has been the most commonly used method for defining cultural heritage and buried remains at archaeological sites. Furthermore, detailed imaging has become an important area of interest [1, 24- 28]. Generally, parallel 2D profile data are acquired in the archaeological site. 3D data imaging, obtained by aligning parallel 2D profile data sets, is used to identify temporal changes at a constant depth. The locations and the depth of the remains in the study area can be determined via slices of the 3D data volume. Therefore, the GPR method gives more precise results than other geophysical methods. However, the obtained results and their interpretation can be further improved when the data set is visualized as a volumetric rendering of the remains. This method allows anyone to imagine how an area looked by looking into the 3D image.

The aim of this chapter is to show transparent 3D GPR data visualization with the most suitable viewing angle into this 3D data volume including buried objects, which is called a transparent 3D half bird's-eye view of the GPR data volume or its sub-volumes. Therefore, first, we introduce the study areas and data acquisition, followed by general data processing steps of the gathered 2D GPR profile data set. Third, we show a revised colour range of the amplitude scale, representing the fourth dimension of the hybrid 2D/3D visualization. Fourth, we attempted to realize a new amplitude–colour-balancing approximation, according to the travel time range or depth range, as an alternative approximation of gain in order to prevent exaggerated amplitude gain, which affects transparency and obscures buried infrastructures. Fifth, we appointed a new opaque function, which must support amplitude–colour scale in order to supply transparency and reveal the fractures, cavities, buried wall remains and foundational infrastructures of interest. The fourth and fifth steps are conducted one within the other.

In this chapter, first, the method was used to show micro-discontinuities in monumental statue groups at Mustafa Kemal ATATÜRK's mausoleum (ANITKABIR) in Ankara, Turkey. Second, the method was used to define buried infrastructures and archaeological remains inside and northeast of the Zeynel Bey tomb, in Hasankeyf ancient city in south-eastern Turkey. The studies examined whether the proposed GPR method could yield useful results at these highly restricted sites. Our visualization results were an interpretation of the two datasets with different objectives.

2. Monitoring micro-discontinuities in monumental statues

2.1. Study area: Mustafa Kemal ATATURK'S Mausoleum, ANITKABIR

The Anitkabir mausoleum, was completed in 1953 in Ankara, Turkey (Figure 1). It hosts state ceremonies during national festivals, and represents the Turkish people and Ghazi Mustafa Kemal ATATÜRK, the founder of the Republic of Turkey (Figure 1b). Anitkabir was constructed in three phases. The first part is an entrance road, called the Lion Road, which is 262m long and has a total of 24 lion statues along each side, representing power and peace (Figure 1c). The second part is a ceremonial square, and the third is the Mausoleum (Figure 1c). At the beginning of the Lion Road, the Turkish people are represented by three large male statues in front of the Freedom Tower on the left side (Figure 1d), and three large female statues in front of the Independence Tower (Figure 1e) on the right side [29].

The monument groups (three women, three men) and twenty-four lion statues of Anitkabir are mainly composed of white travertine from Pinarbasi, Kayseri, Turkey. The white-coloured travertine has a banded and spongy texture under the microscope [23]. It is mainly composed of calcite, aragonite with a small amount of salt, recrystallized calcite, gypsum and plant relicts. Table 1 shows the modal mineralogical composition and physical properties of this travertine. Micro-fractures were observed under a polarizing microscope, especially at the rim of the vesicular of the rocks (Figure 2).

Mineral Composition	Calcite (54%) and Aragonite (31%)
Alteration Products	Recrystallized calcite (13.5) and rarely gypsum and halite (1.5%)
Colour	White
Hardness	3 Mohs, 52.8 Schmidt
Unit Volume Weight	2.52 gr/cm3
Porosity Ratio (%)	9.8 ± 2.180
Cracks Ration (%)	3.4
Alteration Ratio (%)	2.5

Table 1. Mineralogical composition and physical properties of white-coloured travertine in Anitkabir.

Figure 1. (a) Geographical location of the study area, **(b)** Ghazi Mustafa Kemal ATATURK, the founder of the Republic of Turkey, **(c)** the ANITKABIR monument, Ankara-Turkey, **(d)** three large male statues at the beginning of the Lion Road, which is the entrance to Anitkabir, on the left side, **(e)** female statues facing the male group, **(f)** 24 lion statues, representing power and peace, sit on each side of the Lion Road.

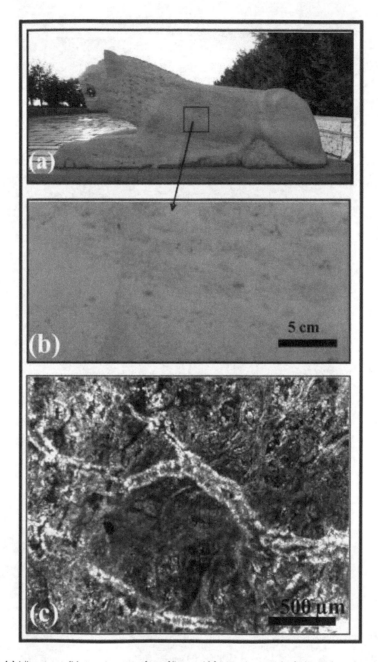

Figure 2. (a) A lion statue, **(b)** view close to surface of lion, and **(c)** microphotograph of white-coloured travertine.

Figure 3. Data measurements on the female statues.

2.2. GPR survey description

The human statues were divided into several subparts such as skirt or under waist, between waist and neck, arms, trousers, etc. for GPR survey. Some parts were divided into additional subparts (Figures 3 and 4) according to their figures, in order to enable true 3D imaging and protect the profile line position, because topography correction was not possible on the statues. In addition, three profiles spaced 10 cm apart were arranged along the backs of lion statues. The data acquisition scheme is shown in Figure 5, in which GPR data profiles gathered on the body of the first female statue were split into two parts, called skirt and upper part between waist and neck.

Profiles were spaced at 10 cm on each subpart, and were lined with a paper band sticker. First, data survey tests were carried out to determine the recording time-window according to the approximate thicknesses of the statues. A RAMAC CU II GPR system equipped with a 1.6-GHz bistatic shielded antenna was employed on all the statue groups. Transmitter–receiver antenna offset was 0.05m. Trace spacing was 0.0044m and time-sampling interval per trace on each profile was 0.0327 ns.

Figure 4. Data acquisitions on the male statues.

2.3. GPR data processing

Data processing was performed on 2D GPR profile data sets for each part of the statues with the REFLEXW program (ver. 5.5) developed by Sandmeier Scientific Software [30]. Start-time correction, then de-wow and background removals were applied to all the profile data in order to protect the true time scale, remove very low frequency effect and average amplitude knowledge respectively. The amplitude decay compensation was applied to all traces of the whole data set by using the same small-scale linear gain function. A second-order band-pass Butterworth filter was used for the whole data set to eliminate low-frequency artifacts and high-frequency noise. The resulting synthetic hyperbolas were matched with diffraction patterns throughout the profiles to determine average velocity of the electromagnetic (EM) wave. The best matching hyperbola provided the velocity of 0.12 m/ns. Finally, Kirchhoff migration was applied to the radargrams using average velocity, in order to carry diffracted electromagnetic (EM) waves true locations.

The quality of GPR images is strongly dependent on appropriate correction of the attenuation effects, usually supplied by time-varying gain. However, historically, the use of amplitude gain in basic processing of GPR data has been highly subjective and also very much displaying methodology [31]. There are various methods available for amplitude gain for GPR data. Traditional time-varying gain is carried out using linear, exponential functions, etc. functions, including ground wave amplitudes. However, this operation is not linear. The time-gained GPR data cannot recover the original information. Selection of the gain function depends mostly on the user and the quality of the GPR data. Both exaggerated linear and exponential time-gain change not only the amplitude range for each time step but also amplitude shape. Exaggerated time gain to image 2D GPR data can result in erroneous interpretation, especially when using such data to construct a 3D volume.

Time-gained signal is

$$s(x,t) = r(x,t)h(t) \tag{1}$$

$$h(t) = a, 1 < a \le \left| \frac{r_{max}}{r_{min}} \right| \tag{2}$$

Where, $h(t)$, is a time gain function, $|r_{max}|$ and $|r_{min}|$ are maximum- and minimum amplitude of the reflected/diffracted wave $r(x, t)$, respectively. $h(t)$, has a constant decimal value a between 1 and 2, and also is a linear function between 1 and a for a 2D data profile [32].

Our approximation concerns 2D or 3D image simplification in determining buried archaeo-logical remains and discontinuities such as fractures. Therefore, we assigned a new amplitude–colour range for 2D radargrams of the GPR profiles and opaque range in order to identify anomalies and display the data set in a transparent 3D data volume. Figure 6a indicates some processed radargrams of the profiles of the skirt of the first female statue, shown in Figure 5.

Figure 5. GPR data acquisition scheme of the first female statue.

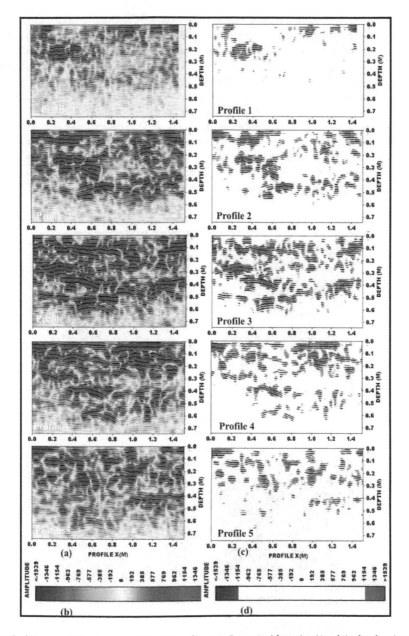

Figure 6. a) Processed 2D radargrams of the GPR profiles 1 to 5, acquired from the skirt of the first female statue, using (b) selected linear amplitude–colour function, (c) the same radargrams using (d) re-arranged amplitude–colour scale of (b) to reveal small cavities and fractures on the radargrams.

A linear colour scale of the radargrams (Figure 6b) indicates the amplitude range, which begins with maximum negative polarity and ends with maximum positive polarity. The blue colour range represents maximum negative amplitude, while the purple colour range represents the positive amplitudes according to their values. It is known that the maximum positive and maximum negative amplitude ranges in the amplitude–colour scale represent the fractures and cavities filled with air or buried archaeological remains in soil. Therefore, it is necessary to check the maximum amplitude ranges on the time slices to identify these target objects. We applied a new approximation to eliminate the weak reflections that are characteristic of cemented rock [7, 23], and to activate the fractures and cavities represented by strong reflections/diffractions on the radargrams (Figure 6c). This involved assigning a new colour scale for the amplitude range of the processed profile data set by means of a new amplitude–colour function rather than linear amplitude–colour function (Figure 6d, Figure 7).

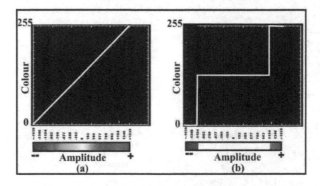

Figure 7. (a) A linear functioned amplitude–colour scale and **(b)** re-functioned colour scale of the same amplitude range.

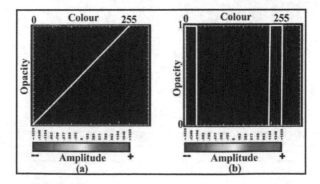

Figure 8. (a) A linear functioned opacity for the linear functioned amplitude–colour scale and **(b)** A re-arranged opacity function using the same amplitude–colour scale to activate only maximum amplitude range and remove all others.

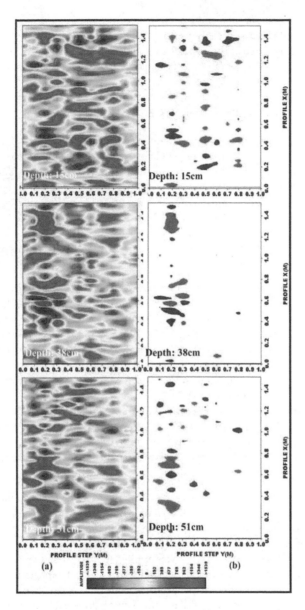

Figure 9. (a) Traditional solid depth slices at 15 cm, 38 cm and 51 cm, **(b)** Transparent depth slices for the skirt of the first female statue (same three depths).

The horizontal axis of the amplitude–colour function (Figure 7) is the amplitude scale of the GPR data, whereas the vertical axis represents colour categories from 0 to 255. The colour limitation allowed simplification and made only fractures and native cavities visible on the radargrams, as seen in Figure 6c. The horizontal x-axis of Figures 6a and 6c indicate the distance along the measuring profile. The vertical axis shows depth range, which represents the thickness of the statue skirt from front surface to back surface, transformed by using average velocity of the EM wave.

2.4. Transparent 3D half bird's-eye view of GPR data set

Generally, interactive visualization is carried out by constructing 3D data volumes of parallel-aligned 2D GPR data sets to show the target objects. The 3D data volume can be displayed as slices, including profiles, times (or depths) and common traces of the profiles; or separated sub-blocks are rendered as solid iso-volumes with linear opacity, determined by the amplitude of the anomalies. The buried fractures or cavities can be defined on the interactive slices, particularly on depth slices with location, and shapes according to depth. Therefore, it was necessary to check the most meaningful depth slices and profiles to define the structures of the subsurface if the area is small and complex. However, the obtained results could be further improved.

Our aim was to obtain a good 3D data volume display, which was a critical part of interpreting the GPR data set. The 3D image is able to present a view of subsurface features such as a fracture or cavity, in addition to objects such as industrial and/or archaeological remains, etc. This imaging could be achieved by a transparent 3D half bird's-eye view revealing only buried objects. Therefore, firstly, transparency could be achieved by constructing an opacity function instead of linear opacity determined by the amplitude scale (Figure 8). The horizontal axis of the opacity function was the amplitude scale starting with maximum negative amplitude and ending with maximum positive amplitude; the vertical axis represented opacity coefficients of the amplitude range [11, 23]. Thus, any amplitude range could be highlighted or minimized by the appointed opacity coefficient. The REFLEXW program allows the opacity coefficient to be chosen between one (maximum opacity) and zero (transparent) (Figure 8b) [26]. A transparent view could be obtained only by eliminating the unwanted amplitude range.

Therefore, the amplitude range was important. Because it was known that the maximum amplitudes represented discontinuities, the weak amplitude range was eliminated by giving these a zero opacity value, and transparent 3D imaging was obtained. The transparency was achieved by allocating an opaque interval to the amplitude scale, similar to the re-arranged amplitude–colour approximation for interested profile range or time range for the solid 3D GPR data volume. This visualization type was applied to both the statues and archaeological remains in this chapter. Figure 9a indicates traditional, solid depth-slices, while Figure 9b indicates transparent depth slices at 15 cm, 38 cm and 51 cm of the data set for the skirt of the first female statue. These slices were used to control micro-fractures and cavities according to the skirt thickness. The horizontal x-axis and y-axis of slices indicate the profile sequence and the distance along the measuring profile respectively. Locations of the micro-cavities could be seen on the slices. It is necessary to carefully check interactive slices in order to determine location and shapes according to depth (thickness).

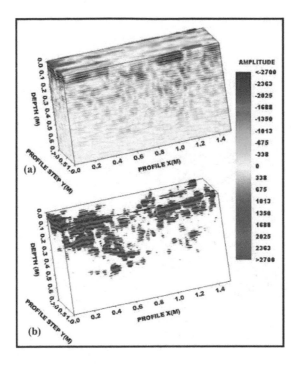

Figure 10. (a) Solid 3D GPR data volume visualization of the skirt of the first female statue (Figure 5) with all profiles and depth range through special view angle of profile slices, **(b)** transparent half bird's-eye views of the same 3D data volume of (a).

Secondly, it was necessary to arrange viewing angles of the x, y and z axes, to obtain the maximum meaningful 3D data volume for the relevant depth range or profile range. The slices could be interpreted differently with differing viewing angles, although the visualization of the slices was required to be the same with the data measurement axes on the map or on a picture. However, there was a general lack of knowledge about the subsurface, including the remains, and about when to take data measurements with regard to information such as the direction of a fracture or an archaeological wall. Therefore, the data measurement strategy was decided according to the field size.

The slices obtained with the same axes on the map of the study site could not effectively represent the subsurface. In addition, when the slices were rotated around the axes, a lining fracture or a wall along the same direction as the angle of view of the slice could be imaged more effectively than in the standard view. To image the fractures and native cavities in the statues, it was decided to use a transparent 3D sub-volume of the profile and depth slices with a half bird's-eye view by arranging the view angles of the axes. Figures 10a and 11a indicate the solid 3D data volume visualization of the skirt of the first female statue with all profiles and depth range using a special viewing angle of profiles and depth slices. In addition, Figures

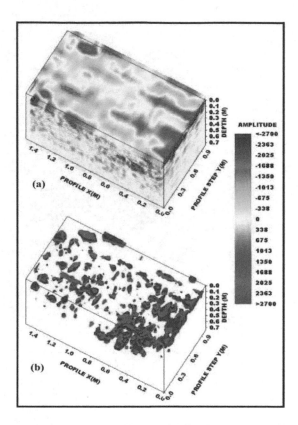

Figure 11. (a) Solid 3D data volume visualization of the skirt of the first female statue (Figure 5) with all profiles and depth range through special viewing angle of depth slices, **(b)** transparent half bird's-eye visualization of the same 3D data volume in (a).

10b and 11b represent our approximation with the transparent 3D half bird's-eye view visualization of the same data set. The horizontal x-axis of Figure 10 and Figure 11 indicates the distance along the profile. The horizontal y-axis represents the profile sequence. The vertical axis indicates thicknesses of statues from the front surface to the back surface of the skirt of the statue. Figure 12 shows different depth ranges of transparent 3D half bird's-eye views of the GPR data aligned on the first female statue (Figures 5 and 11b) between 0–10 cm, 10–20 cm, 20–30 cm 30–40 cm, 40–50 cm and 50–60 cm depth ranges, and shows the locations of the micro-fractures and cavities with purple and blue colours, represent the maximum amplitude ranges. Figure 13 indicates different profile ranges of transparent 3D half bird's-eye views of the GPR data aligned on the skirt of the first statue (Figures 5 and 10b) between profiles 1–3, profiles 4–6, profiles 7–9 and profiles 10–11; and the upper part of the female statue between profiles 1–3 and profiles 4–8 through special viewing angle of the profile slices; the locations of micro-fractures and cavities are represented by purple and blue colours.

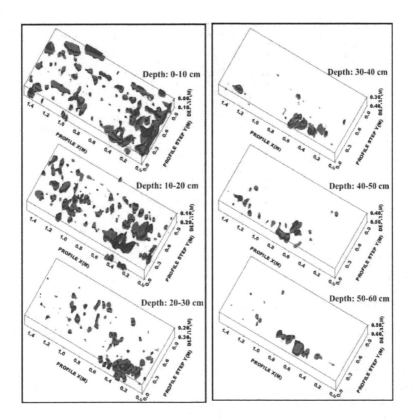

Figure 12. Transparent half bird's-eye view results of the 3D sub-volumes of Figure 11b between 0–10 cm, 10–20 cm, 20–30 cm, 30–40 cm, 40–50 cm and 50–60 cm depth ranges, respectively, including internal micro-fractures and native cavities.

The profile ranges of Figure 13 give the locations of the fractures and cavities throughout the depth of the profile ranges, while the depth ranges of Figure 12 give the locations of the fractures and cavities along the full surface of the skirt through the depth ranges. Therefore, it is possible to check both transparent profile ranges and depth ranges according to the most appropriate viewing angle of the 3D GPR data volume in order to determine the locations and

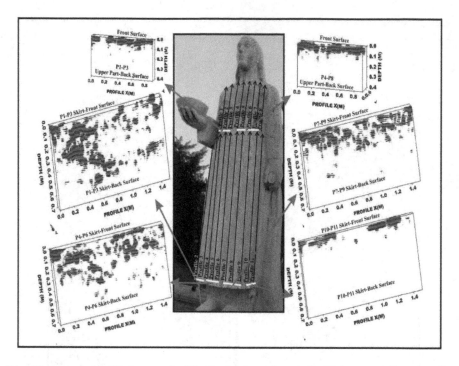

Figure 13. Transparent half bird's-eye results of the 3D sub-volumes shown in Fig. 10b between profiles 1–3, profiles 4–6, profiles 7–9 and profiles 10–11; and the upper part between profiles 1–3 and profiles 4–8 through special viewing angle of profile slices, including internal micro-fractures and cavities.

continuities of micro-fractures and cavities within very restricted study areas, such as those in the present study.

The native micro-cavities are not effective to harm the lions. The micro-fractures show a lateral, inclined or vertical linearity. According to Figures 12 and 13, the skirt of the statue had an important fracture between profiles 1 and 3, ranging from 0–80 cm from the front surface as far as the back surface. In addition, the figure had more small fractures and native cavities between 0- and 30-cm depth. To summarize to our method, we present results from the transparent 3D half bird's-eye view of three GPR data sets gathered form the backs of three lion sculptures, using three parallel-aligned profiles along the leg and head to visualize interior fractures (Figure 14).

According to the visualization results in the first lion, there were three large fractures aligned parallel to the surface along the back legs, which continue from the upper surface of the back legs to the border lion along the depth; and there were native cavities in the back through the belly. The second lion mostly had disorderly native cavities and micro-fractures along the back surface until the belly. The last lion was seen very powerful, and only included some native cavities in the back side.

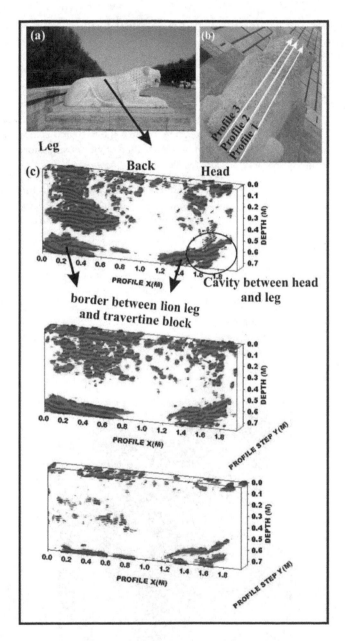

Figure 14. (a) One of the 24 lions on the Lion Road, **(b)** data acquired from three parallel profiles from the back to the head of the lion, **(c)** the results of the transparent half bird's-eye view of the three different lions.

3. Picturing buried archaeological remains and foundational infrastructures

3.1. Zeynel Bey tomb in Hasankeyf ancient city

Our new 3D visualization was applied to archaeological remains both inside and outside the Zeynel Bey tomb in the ancient Turkish city of Hasankeyf. This site is among the last remaining locations of the Silk Road in Anatolia, spreading towards the East, and is located in Batman province, southeastern Turkey (Figure 15). A similar type of visualization for archaeological remains was introduced by previous studies [26, 33].

The precise foundation date of Hasankeyf is not known. The geopolitical situation in Hasankeyf strengthens the possibility of its being a very ancient settlement area. Hasankeyf is identified with the tomb built by Uzun Hasan for his son Zeynel Bey, who died in the war of Otlukbeli (1473) by the Tigris [34]. The Zeynel Bey tomb, the first example of the Anatolian mausoleum tradition (Figure 16), is on the north bank of the Tigris, across from the city.

The tomb is a cylinder of diagonal patterns made using brick and tile, with a pointed arch portal doorway on the north and a window in the south wall (Figure 16). Above the main shaft is a slightly smaller diameter shaft, which has small windows in each of the cardinal directions and carries a hemispherical dome (Figures 16 and 17) [34-37]. Inside, the plan is octagonal, with muqarnas niches supporting the transition to the round base of the dome. Each of the eight walls has a rectangular arched niche, and the burial chamber is recessed into the floor (Figure 17) [36].

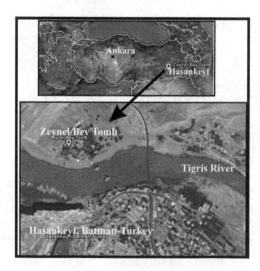

Figure 15. Geographical map of the Zeynel Bey tomb in Hasankeyf ancient city, Turkey.

Figure 16. Appearance of the Zeynel Bey Tomb, the first example of Anatolian mausoleum tradition, with some architectural details.

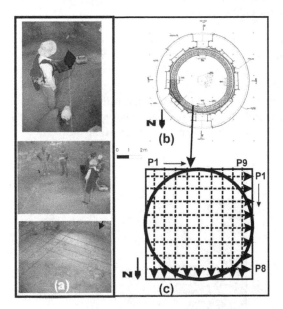

Figure 17. (a) GPR data measurements inside the Zeynel Bey Tomb, **(b)** Interior plan of Zeynel Bey Tomb: octagonal, with muqarnas niches supporting the transition to the round base of the dome, **(c)** The data measurement plan inside the tomb.

Hasankeyf and many other Tigris Valley settlements that have previously directed world history will be submerged when the proposed Ilisu Dam is completed. Despite proposals to move the monuments, many historic sites and artifacts will be lost when the reservoir is filled. Efforts to relocate and preserve culturally significant sites are currently led by Prof. Dr. Abdüsselam Uluçam, the rector of Batman University [37].

3.2. GPR data measurements at the Zeynel Bey tomb

In this section, we present only two parts of the huge study area, including the tomb of Zeynel Bey and the Ottoman bath, an area of approximately 150×200m. The first part was inside of the tomb. The Zeynel Bey tomb is 4m along the east–west orientation and 3.5 m along the south–west (Figure 17). A RAMAC CUII GPR system was used with a bi-static 500-MHz center band shielded antenna to acquire the profile data. Within the tomb, 9 parallel profiles spaced 0.5m apart were directed from east to west, and 8 profiles were directed from north to south, making a total of 17 profiles (Figure 17c). The second part of the survey was conducted on the northeast side of the tomb, and 19 profiles were directed from south to north on the east side (Figure 18). Parallel profiles were spaced 1m apart, and each profile had a trace spacing of 5cm and a 70ns time-window per trace.

3.3. Data processing and a new amplitude-balancing approximation for transparent 3D half bird's-eye view of the GPR data set

The GPR data, gathered within and on the northeast side of the tomb, were processed using REFLEXW software. After sequencing the profiles as discussed at the Anitkabir site, the start-time correction was applied. De-wow and background removals were applied. The second-order band-pass Butterworth filter was then applied to the whole data set, to eliminate unwanted frequency noise. A simple linear gain function was applied as discussed in section 2.3. Velocity analysis indicated that the average velocity of electromagnetic wave propagation was 0.11m/ns. Finally, Kirchhoff migration was applied to the data.

Figure 18. (a) GPR data measurement on the northeast side of the Zeynel Bey tomb.

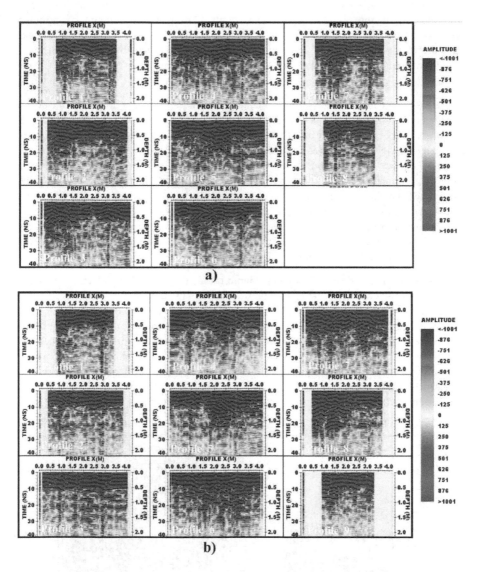

Figure 19. Processed radargrams of the profiles along (a) the east–west direction and (b) the south–north direction inside the Zeynel Bey tomb.

The remains of a buried archaeological wall and foundational infrastructures can be defined on depth slices with location, and shapes according to depth. However, depth slices may not explain the subsurface if the area is small and complex, as in the Zeynel Bey tomb. Therefore, it is necessary to check the most meaningful depth slices and profiles to define the subsurface structures. During the control of the interactive depth slices, it is also important to control the colour of the amplitude range of the remains. It is known that the maximum amplitude ranges on an amplitude–colour scale represent subsurface features. However, the amplitude decreases with increasing recording time even if an appropriate time gain is applied to the profile data.

It is also known that if the colours of maximum amplitude ranges are dominant at the beginning of the recording time or at depths very near to the surface, then the resolution of the slices can not be sufficiently high to differentiate buried structures very near to the surface, especially for small study areas as seen on Figure 19. Our third approximations after re-arranged opacity function and viewing angle into the 3D volume, related to achieving amplitude balance according to the recording time or the depth (Figure 20). The aim of the amplitude balancing was to protect the colour range of the anomaly representing the target remains according to the time or depth range of the 3D sub-volumes; this was achieved by increasing or decreasing the maximum amplitude values of the amplitude–colour scale for the special depth slice or for a depth range. Therefore, the full data volume should be divided into sub-data volumes in time or depth.

To balance the amplitude–colour scale in a depth range, firstly the changing colour range of the remains was controlled. Then, the amplitude range of the remains was determined in the 3D sub-volume and the maximum amplitude values of the colour range were reduced or increased in the depth range. The balancing of the amplitude–colour scale according to the amplitude range of the buried wall remnants produced better resolution and showed the remains with the same colour ranges on the slices or their 3D sub-volumes with increasing time or depth axis. The balancing procedure was also a time gain, which was a non-uniform stair function, weighting the electromagnetic wave field according to the time axis (Figure 20). The approximation was important to obtain a 3D representation of the volume.

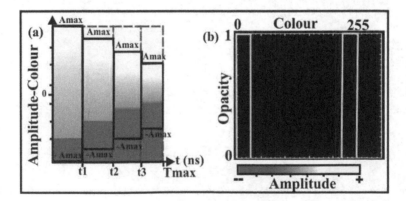

Figure 20. (a) Re-scaling maximum amplitudes according to selected time or depth range of 3D GPR sub-volumes, and assigning the same colour range; **(b)** assigning a new opaque range according to the re-scaled amplitude–colour range.

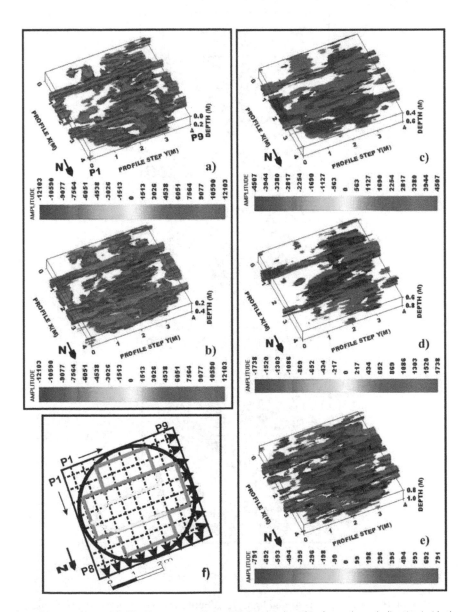

Figure 21. Transparent 3D half bird's-eye views of the GPR data set aligned in the south–north direction inside the Zeynel Bey Tomb between **(a)** 0 and 20 cm, **(b)** 20 and 40 cm, **(c)** 40 and 60 cm, **(d)** 60 and 80 cm, and **(e)** 80 and 100 cm depth ranges; and **(f)** infrastructures (orange colour) of the base and a cemetery (yellow colour) in the base wall inside the tomb.

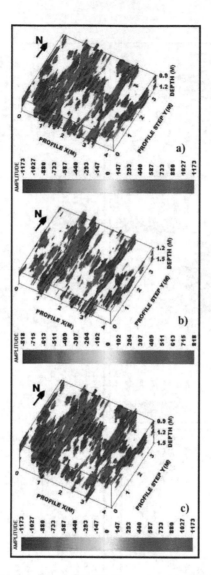

Figure 22. Transparent 3D half bird's-eye views of the GPR data set aligned east–west inside the Zeynel Bey tomb at depths between (a) 90 and 120 cm, (b) 120 and 150 cm, and (c) 90 and 150 cm.

Practically, it was only possible to highlight the amplitude–colour ranges representing the remains, which were previously established as maximum amplitude–colour ranges of the 3D sub-data volumes. It was therefore straightforward to construct an opacity function by appointing opacity coefficients of one (maximum opacity) for maximum negative and maximum positive amplitude–colour ranges and zero (transparent) for unwanted amplitude–colour ranges on the amplitude–colour scale. A transparent 3D half bird's-eye view was obtained only by eliminating unwanted amplitude range to reveal subsurface remnants and features. However, depth range depended on the maximum amplitude volume. In order to construct a large depth range, the maximum amplitude–colour range would need to be restricted more than the normal range to produce a meaningful image. Figure 21 shows transparent 3D sub-data volumes with half bird's-eye view between 0 and 20cm, 20 and 40 cm, 60 and 80 cm, and 80 and 100cm depth ranges from the GPR data set aligned in the south–north direction inside the tomb. Wall structures and a cemetery with stair could be clearly seen. In addition, Figure 22 shows transparent 3D sub-data volumes between 90–120 cm. 120–150 cm and 90–150cm depth ranges along the east–west direction inside the tomb. The visualization results of the east–west side data set supported those of the south–north. However, the results of the south–west side perfectly imaged the foundational infrastructure and buried cemetery.

Figure 23. (a) The North side of the Zeynel Bey tomb, **(b)** the excavation on (a) and a cemetery, (c) the walls of the cemetery and stairway.

Although the maximum diameter inside the tomb was 4 m, it would be possible for users to envisage buried archaeological remains when viewed using the transparent 3D image. Figure 23 shows the excavation results inside the tomb. The excavation findings perfectly matched the transparent 3D half bird's-eye view. The visualization results also precisely indicated all details of the foundational wall remains and the cemetery with stairs.

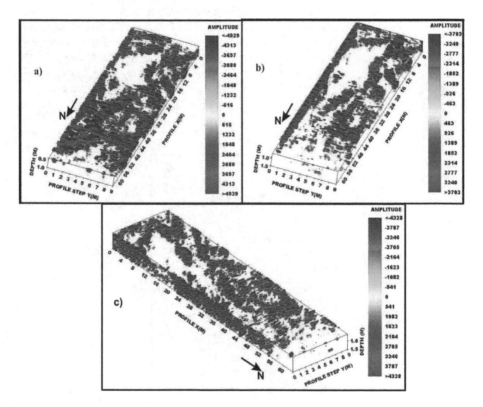

Figure 24. Transparent 3D half bird's-eye views of the GPR data set on the northeast side of the Zeynel Bey tomb, and pictured buried wall remains in depth between **(a)** 50 and 100 cm, **(b)** 75 and 150 cm; **(c)** 75 and 150 cm but with different viewing angles of the x, y and z axes.

Half bird's-eye views of the transparent 3D depth volume ranges of the GPR data were also produced for the northeast side of the tomb (Figure 24). The wall structures were seen exactly in the corresponding transparent 3D imaging. The results showed that the buried walls were very near the surface and very complex. These imaging results remain to be confirmed through direct observation, as excavation has not yet taken place at these sites. The results could be confirmed by comparing the excavations inside of the tomb and their 3D visualization results.

4. Results and conclusions

Interactive transparent 3D visualizations of GPR data volumes were produced for each part of 6 human statues and a group of 24 lion sculptures at Anitkabir, Turkey. We produced images of natural internal cavities and fractures within the statues via a proposed method of interactive transparent 3D half bird's-eye visualization of the GPR data volume of the profile range and the depth range.

Our first approximation related to amplitude–colour simplification. This allowed us to differentiate and locate native micro-cavities and micro-fractures inside the statues, based on radargrams of the profiles gathered on the surface of the statues. The second approximation related to our proposed opacity function, which dominated maximum positive and negative amplitudes and eliminated other irrelevant amplitudes. Third, an interactive transparent 3D half bird's-eye view of the 2D GPR data set was achieved by carefully assigning the amplitude–colour scale and its opacity range, together with a carefully selected viewing angle, profile and/or depth range. The transparent 3D imaging proved successful in identifying changes in the statues, accurate x–y locations and accurate depths. This monitoring replied to the aim of the study. Mapping fractures and cavities within statue groups could enable evaluation of their stability and indicate the best way to minimize restoration costs.

In the fourth approximation, the present study developed an improved amplitude-balancing approximation in order to reveal and differentiate subsurface historical remains from the surrounding soil medium. By combining the second and third approximations, the balancing of the amplitude–colour scale achieved sufficiently high resolution to represent the remains with the same colour ranges on both the slices and their transparent 3D imaging with increasing depth axis. A viewing angle was allocated to the x, y and z axes of the 3D data-volume by checking wall orientations and the data measurement directions on the map to obtain a good half bird's-eye view. The transparent 3D half bird's-eye view of the 2D GPR data set provided better imaging to accurately visualize the subsurface by sensing x–y locations and depths. The results demonstrated that the GPR method and our developed 3D visualization gave perfect results in a closed, circular area including remnants of very complex buried foundational wall, including a cemetery, within the Zeynel Bey tomb without any risk of damage to the tomb and infrastructures. In addition, the results also indicated that the visualization method perfectly monitored the archaeological buried walls with accurate x–y locations and accurate depths.

We also indicated that GPR method provides highly accurate results for the position and depth of targets within very complex and restricted areas, even on statues.

Acknowledgements

The authors would like to thank the military authorities at ANITKABIR; KA.BA Conservation of Historic Buildings and Architecture Ltd. with the coordination of rescue excavation director Prof. Dr. Abdüsselam Uluçam, Rector of Batman University, Turkey; and Ankara University

Earth Sciences Application and Research Center (YEBIM) for supporting the projects. The authors also thank Prof. Yusuf Kagan Kadioglu, head of the Geological Engineering Dept., Ankara University, Turkey; Asst. Prof. Ali Akin Akyol, Baskent Vocational Higher School Program of Restoration and Conservation, Ankara University; and the MSc student group of the Ankara University Geophysical and Geological Engineering Departments for helping us in data collection at Anitkabir.

Author details

Selma Kadioglu[1,2*]

Address all correspondence to: kadioglu@ankara.edu.tr

1 Ankara University, Faculty of Engineering, Department of Geophysical Engineering, Ankara, Turkey

2 Ankara University, Earth Sciences Application and Research Center, Ankara, Turkey

References

[1] Daniels JJ. Ground Penetrating Radar for Imaging Archaeological Objects in the Subsurface: Proceedings of the New Millennium International Forum on Consideration of Cultural Property, Kongju, Korea. 2000; 247-265.

[2] Daniels DJ. Ground Penetrating Radar. The Institution of Electrical Engineers. Second Edition, London, United Kingdom, 2004.

[3] Davis JL, Annan AP. Ground-Penetrating Radar for High Resolution Mapping of Soil and Rock Stratigraphy. Geophysical Prospecting 1989; 37 531–555.

[4] Grandjean G, Gourry JC. GPR Data Processing for 3D Fracture Mapping in a Marble Quarry (Thassos, Greece). Journal of Applied Geophysics 1999; 36 19–30.

[5] Orlando L. Ground Penetrating Radar in Massive Rock: A Case History. European Journal of Environmental and Engineering Geophysics 2002; 7 265–279.

[6] Grasmueck M. 3-D Ground Penetrating Radar Applied to Fracture Imaging in Gneiss. Geophysics 1996; 61 (4) 1050–1064.

[7] Grasmueck M, Weger R, and Horstmeyer H. Full-Resolution 3D GPR Imaging. Geophysics 2005; 70 (1) K12-K19.

[8] Tsoflias GP, Gestel J-PV, Stoffa PL, Blankenship DD, Sen M. Vertical Fracture Detection By Exploiting the Polarization Properties of Ground-Penetrating Radar Signals. Geophysics 2004; 69 (3) 803–810.

[9] Kadioglu S. Photographing Layer Thicknesses and Discontinuities in a Marble Quarry with 3D GPR Visualisation, Journal of Applied Geophysics 2008; 64 109-114.

[10] Porsani JL and Sauck WA. GPR Profiles Over Multiple Steel Tanks:Artifact Removal Through Effective Data Processing. Geophysics 2007; 72(6) J77-J83.

[11] Kadioglu S and Daniels JJ. 3D Visualization of Integrated Ground Penetrating Radar Data and EM-61 Data to Determine Buried Objects and Their Characteristics. Journal of Geophysics and Engineering 2008; 5 448-456.

[12] Khadr N, Barrow BJ and Bell TH. Target Shape Classification Using Electromagnetic İnduction Sensor Data *Proc. UXO Forum'98* Online at http://citeseer.ist.psu.edu/khadr98target.html

[13] Pasapane BE and Sieling DR. Utilizing A Multiple Metal Detector Array for Locating Anomalies Inductive of Ferrous and Non-Ferrous OE/UXO 2002; 02P-0157, 28th Environmental &Energy Symposium & Exhibition (Ray F. Weston Inc., Charleston, South Carolina); 2002.

[14] Won J, Keiswetter D and Bell T. Electromagnetic Induction Spectroscopy for Clearing Landmines. IEEE Transactions On Geosciences And Remote Sensing 2001; 39 703-709.

[15] Benson A K Applications of Ground Penetrating Radar In Assessing Some Geological Hazards: Examples of Groundwater Contaminants, Faults, Cavities J. Appl. Geophys. 1995; 33 177–93.

[16] Yoder RE, Freeland RS, Ammons JT and Leonard LL. Mapping Agricultural Fields with GPR and EMI to Identify Offsite Movement of Agrochemicals. *J. Appl. Geophys* 2001; 47 251–9.

[17] Stroh JC, Archer S, Doolittle JA and Wilding L. Detection of Edaphic Discontinuities with Ground-Penetrating Radar and Electromagnetic Induction. *Landscape Ecol.* 2001; 16 377–90.

[18] Koralay T, Kadioglu S And Kadioglu YK. A New Approximation in Determination of Zonation Boundaries of Ignimbrite by Ground Penetrating Radar: Kayseri, Central Anotalia, Turkey. Environ. Geol. 2007; 52 1387–97.

[19] Kofman, L, Ronen, A, Frydman, S. Detection of Model Voids by Identifying Reverberation Phenomena in GPR Records. Journal of Applied Geophysics 2006; 59 284–299.

[20] Orlando L and Slob E. Using Multicomponent GPR to Monitor Cracks in a Historical Building J. Appl. Geophys. 2009; 67 327–34.

[21] Masini N, Persico A, Guide A and Pagliuca A. A Multifrequency and Multisensory Approach for the Study and the Restoration of Monuments: The Case of the Cathedral of Matera. Advances in Geosciences 2008; 19 17–22.

[22] Bavusi M, Di Napoli R and Soldovieri F. Microwave Tomographic Approach for Masonry Investigation: Some Real Results. Advances in Geoscience 2010; 24, 83–8.

[23] Kadioglu S and Kadioglu, YK. Picturing Internal Fractures of Historical Statues Using Ground Penetrating Radar Method. Advances in Geosciences 2010; 24 23-34, www.adv-geosci.net/24/23/2010/.

[24] Leucci G and Negri S Use of Ground Penetrating Radar To Map Subsurface Archaeological Features İn An Urban Area. J. Archaeological Sci. 2006; 33, 502–12.

[25] Masini N, Nuzzo L and Rizzo E. GPR Investigations for the Study and the Restoration of the Rose Window of Troia Cathedral (Southern Italy) Near Surf. Geophys. 2007; 5 287–300.

[26] Kadioglu S. Definition of Buried Archaeological Remains with a New 3D Visualization Technique of Ground Penetrating Radar Data Set in Temple Augustus in Ankara, Turkey. Near Surf. Geophys. (Special issue on GPR in Archaeology) 2010; 8 397–406.

[27] Goodman D Ground penetrating radar simulation in engineering and archaeology. Geophysics 1994; 59 224–232.

[28] Goodman D, Schneider K, Piro S, Nishmura Y and Pantel AG. Ground penetrating radar advances in subsurfaces imaging for archaeology. In: Remote Sensing in Archaeology (eds J. Wiseman and F. El-Baz), 2007; 367–386, Springer.

[29] http://www.tsk.tr/eng/anitkabir: website of the Turkish General Staff (TSK).

[30] Reflexw. Sandmeier Scientific Software 2007

[31] Annan AP, Practical processing of GPR data. Proceedings of Second Government Workshop on Ground Renetrating Radar, Sensor and Software Inc. 1999.

[32] Kadıoğlu S, Daniels JJ. Different Time Gain and Amplitude-Color Arranging for Ground Penetrating Radar Data: Applied Samples: XIII International Conference on Ground Penetrating Radar Lecce-ITALY, 2010. IEEE Xplore Digital Library Conference Puplications, Ground Penetrating Radar (GPR), 2010; 13th International Conference, E-ISBN : 978-1-4244-4605-6; DOI: 10.1109/ICGPR.2010.5550165, 2010.

[33] Kadioglu S, Kadioglu YK and Akyol AA. Monitoring Buried Remains with Transparent 3D Half Bird's-eye View of Ground Penetrating Radar Data in the Zeynel Bey Tomb In the Ancient City of Hasankeyf - Turkey, J. Geophys. Eng. (Special Issue on Cultural Heritage) 2011; 8 S61-S75.

[34] Aslanapa O. Turkish Art and Architecture. New York: Praeger, 1971.

[35] Basgelen N. A Unique Global Heritage Sstarting to Count its Remaining Days: Hasankeyf and Tigris Valley. Archaeology / Monument-Environment (Arkeoloji /Anit-Çevre) 2006; September-October 17 p114-119.

[36] Eskici B, Akyol AA and Kadioglu YK. Material Analyses and Conservation Problems of the Hasankeyf Zeynel Bey Tomb. Journal of Turkish Archaeology and Ethnography 2008; 8, 15-37. (in Turkish)

[37] Uluçam. A. Excavations of Hasankeyf from the Past to Nowadays, Spatial Issue of Konya Book, December 2007. p.681-710. (in Turkish)

Radioanalytical Techniques: Cases Studies and Specific Applications of NAA

Chapter 6

Nuclear Analytical Techniques in Animal Sciences: New Approaches and Outcomes

A.C. Avelar, W.M. Ferreira and M.A.B.C. Menezes

Additional information is available at the end of the chapter

1. Introduction

"Interest in problems relating to the food and nutrition of man is already widespread and sincere (...) The time is not distant when

it will be generally recognized that man should pay at least much attention to problems relating to his own food as to the study of the

food of domestic animals". [1]

Those statements above were made by the US Secretary of Agriculture, J. Sterling Morton in the name of the USDA (United States Department of Agriculture) in 1896 to introduce the publication '*The Chemical Composition of American Food Materials*' of Atwater and Woods [1].

In his Letter of Transmittal, 116 years ago, Sterling Morton pointed-out the relevance of the knowledge of nutritive values of national food materials, since there were available in North American only results made in German products [1].

Since 1896, sensitive analytical techniques have become available which allow measurement of essential and toxic elements in food, feed and animal products. Several improvements have taken place in the last decades contributing to great improvements in the analytical quality of results produces. Substantial progress was also achieved by providing standardized equipment based in semiconductor detectors.

Chemical analyses have been improved and laboratories are largely widespread in the five continents, but Sterling's concerns are contemporary and vivid as never before.

In 1936, Hevesy and Levi first utilized a neutron source to analyze dysprosium in Y_2O_3 inaugurating an era of great development in studies involving multi-element determination in many and varied areas of Sciences, including: material engineering, chemistry, agronomy, animal sciences and nutrition. Later, the Hevesy and Levi's technique were identified as Neutron Activation Analysis [2].

The Instrumental Neutron Activation Analysis (NAA) has a great advantage to other (humid) techniques due to the absence of effects of chemical binding to the trueness of results since the NAA is the only technique for quantitative element determination based on phenomena occurring in the atomic nucleus. The total dissolution may not be guaranteed for the entire sample in humid techniques [3].

Other outstanding characteristics include: element specificity, multi-element determination capacity, and sensitivity [2,3].

Potential interferences, sources of error and contributions to uncertainty of measurement are well known and quantifiable [3].

Some authors define NAA as "Mature, completed in development", a stage in which the initial problems have been overcome. However, there are various analytical challenges in the many applications for which INAA may be the preferred technique to obtain information on elements and their concentrations [3].

Oppositely, applying others analytical techniques to study solid samples by flame atomic absorption spectroscopy, graphite furnace absorption spectroscopy, inductively coupled plasma spectroscopy, or inductively coupled plasma mass spectrometry, the sample must first be digested to get the analyte metals in solution. Digestion dissolves only those fractions of metals that can be put into solution under relatively extreme conditions and therefore enables measurement of available metals. Sample digestion by humid procedures generally uses highly corrosive reagents that are strong acids and strong oxidants and demand expert personnel using the proper equipment, including fume hoods and adequate personnel protection. It is expected new developments of high-powered microwave digestions systems coupled to these "open-sample dependent" techniques [4].

2. Neutron activation analysis in Brazil in the new millennium

Brazil has developed technology based on nuclear research reactors, two (IPEN and CDTN) work on neutron activation analysis. At IPEN, São Paulo, there is one research reactor (a 5 MW pool type) and a cyclotron both involved with radioisotopes production. At the CDTN, Belo Horizonte, there is a very active Triga research (IPR-R1) reactor [5]. Two unique duties in both of these institutes are promoting of basic nuclear teaching for students and workers in nuclear industry and embracing the applied research in many fields like: medicine, nutrition, animal sciences, geology, environmental sciences among others [5].

There is an interesting study of lunch meals served at the Cafeteria of the School of Public Health used by students and workers in the University of Sao Paulo (USP). That study could be considered an turning-point in Nuclear Analytical Applications in Brazil, giving pace for a new generation of studies in Life Sciences. The aim of that study from 2000 was to assessing the nutritional adequacy of diets served to University students related to essential elements and also monitoring for some toxic elements [6].

Since 2000, our research group has been dedicated to apply Neutron Activation Analysis in Animal Sciences. Starting from acquiring data for minerals in bovine tissues of the Brazilian cattle; at that point we could not find any study involving such tissues in Brazil in the nutrition field [7].

In that first study [7], our objective was focused to assess the elemental composition of animal tissues to support health and nutritional studies. Since cattle were and still are the most prevalent source of protein among Brazilian families, including meat, milk and dairy products. Determining the elemental concentration in cattle tissues is especially important because these materials are used for multipurpose objectives such as the assessment of animal health, the quality of human foods consumed, and as a potential environmental biomonitor. Chromium, copper, sodium, potassium, iron, and zinc levels were determined in bovine tissues—kidney, liver and muscle—from cattle bred and raised in a potentially metal contaminated region because of mineral activities.

We verified the essential element concentration and possible contamination by toxic elements in cattle tissues that can affect human nutrition. There was a good agreement the between values reported international organizations such USDA and FAO and the Brazilian analytical results obtained; the required data quality was also achieved. The higher iron, chromium, and copper concentrations could reflect the influence of the fate of environmental contaminants depositing in the animal tissues by biochemical processes. The pollutants in the environment reaching the livestock through water and forage may have caused it. In this first study it is not possible to affirm that the higher iron, chromium, and copper concentrations mean that these elements are playing the role of toxic elements because it was a preliminary sampling. However, the presence of such elements that is not reported elsewhere should be verified in detail during other studies. It is important to analyze animal tissues since this matrix can be used an efficient biomonitor to assess the animal's health, and the quality of human foods as well. The application of k0-instrumental neutron activation analysis was considered an effective multi-elemental method used to determine mineral concentration of biological material.

It is well established that (tropical) Brazilian soils are phosphorus deficient or/and have low phosphorus bioavailability as long most of the phosphate molecules are biding in insoluble composts [8]. This phosphorus deficiency affects both crops and livestock produced in Brazil. To avoid it, it has been applied phosphorus sources in soil, plant and animal nutrition. The most popular sources of phosphorus are rock phosphates, dicalcic phosphate, and in some cases, bone meal.

Rock phosphates are a source of many elements considered as contaminants in animal nutrition and to the environment as well. These product are plenty used in Animal Nutrition.

3. A Case of nuclear analytical application in animal sciences: Uranium in rock phosphates and rabbit muscles from animals receiving uranium determined by neutron activation analysis

Nuclear applications in Animal Sciences are not a novelty. It could be considered as the two first studies from 1949 using rabbits to assess the biological consequences of the uranium ingestion: Decreased body weight was reported for rabbits exposed to 11 mg U/m^3 as uranium tetrachloride dust for 35–40 days [9]. Rabbits lost 22% of their body weight during a 30 days exposure to 0.9 mg U/m^3, dogs and cats lost approximately 25% of their body weight during a similar exposure to 9.5 mg U/m^3. Similar effects were observed with uranium tetrafluoride [10].

Introduction: Phosphorus (P) deficiency in crops is an important constraining factor n plant and animal yields, especially in hot humid tropics where soils are predominantly acidic and often extremely P deficient with high P fixation capacities [11,12].

Phosphorus combines with oxygen forming oxides called phosphates. Phosphates are defined as compounds, which contain phosphorus-oxygen (P-O) linkages. The P-O bond has a length of 1.62Å with bond angles of 130° at the oxygen atoms and 102° at the phosphorus atoms at the pentoxide P_2O_5, the only oxide of phosphorus that is produced commercially [13].

Phosphate rock denotes the product obtained from the mining and subsequent metallurgical processing of phosphorus bearing ores. By flotation of phosphate rock it is formed apatite concentrates. These phosphate products are the major phosphorus sources in soil nutrition, also plenty used in animal formulations by industries [14].

...'On their way up the chimney the gases go through four separate treatments. P_2O_5 used to go right out of circulation every time they cremated some one. Now they recover over ninety-eight per cent of it. More than a kilo and a half per adult corpse. Which makes the best part of four hundred tons of phosphorus every year from England alone." Henry spoke with a happy pride, rejoicing wholeheartedly in the achievement, as though it had been his own. "Fine to think we can go on being socially useful even after we're dead. Making plants grow.'...] Brave New World, Aldous Leonard Huxley [15].

This text above was extracted from the 1932 fictional worldwide bestseller *Brave New World* [15]. It could sound so fantasist back when it was firstly published. Maybe it will sound reality in a future not far ahead. Currently, a reduced and shortening number of phosphate sites are mined around the world. Based on current phosphate extraction rates and economic trends in the 1990s, more than half of main phosphate producer countries will have exceeded the life of their reserves up to 2024 [16].

Phosphate ranks second (coal and hydrocarbons excluded) in terms of gross tonnage and volume of international trade [16].

Sedimentary ore deposits have provided about 80%–90% of world production of phosphate [6]. Oppositely, in Brazil igneous deposits represent 80% of the national reserves of phosphate rocks. The major Brazilian phosphate rock reserves are concentrated in the states of Minas Gerais, Goiás and São Paulo [17].

In 2009, the Brazilian phosphate-related products market was shared as phosphate fertilizers (89.12%), animal feed – mainly dicalcium phosphate (6.91%), soil amendment (1.01%), and the remaining was not informed (2.96%) [17].

Ingredient (tons.year^{-1})	2005	2011	2012$^{\triangledown}$
Dicalcium Phosphate	216.400	404.761	560.274
Bone Meal	225.400	n.i.	n.i.
Limestone	634.000	1.247.016	1.276.388

n.i., not informed., $^{\triangledown}$ forecasted in 2012, May for the entire year

Table 1. Phosphorus and calcium sources used in the Brazilian Feed Industry [8]

Products	Quantity (tons)	FOB x U$1,000.00
Raw material	915.449	84.040
Industrialized	2.861.719	945.170
Total	3.777.168	1.029.210

Table 2. Brazilian imports of phosphates, raw and industrialized products [17]

Country	Raw Material (%)	Industrialized Material (%)	Total (%)
United States	n.a.	33.31	30.59
Algeria	20.90	n.a.	n.a.
Morocco	63.80	24.54	27.74
Israel	10.30	8.97	9.08
Tunisia	4.71	7.95	7.68
Russia	n.a.	6.74	6.19
Spain	0.13	n.a.	n.a.
Others	0.16	18.49	18.71

Table 3. Brazilian imports of phosphates by countries, raw and industrialized products. Total pondered by the price FOB, DNPM [17]

Data in table 1 demonstrate the economic relevance of the products under consideration in this study for the Brazilian feed industry. Foreign products are also of concern, as long Brazilian crops and animal yields rely on a great amount of imported phosphate products (tables 2 and 3).

Indeed, any phosphate mined worldwide may contain accessory-gangue minerals and impurities that can be hazardous to man and animal such: Cd, Hg, Pb and V [16].

The United States Agency for Toxic Substances and Disease Registry [19] publishes the CERCLA Priority List of Hazardous Substances that includes substances, which have been determined to be of the greatest public health concern [19]. Uranium is the 97th substance ranked in the list (table 4).

Rank	Substance	Total Points*
1	Arsenic	1665
5	Polychlorinated Biphenyls	1344
97	Uranium	832

*The ranking of hazardous substances on the CERCLA Priority List is based on three criteria (i,ii,iii). They form altogether the Total Score = Σ (i) Frequency of Occurrence + (ii) Toxicity + (iii) Potential for Human Exposure = Σ (i) up to 1.800 Points + (ii) up to 600 points + (iii) up to [300 concentration points + 300 exposure points] [20]

Table 4. Compilation of some hazardous substances (including uranium) in the CERCLA List [19, 20]

This study deals with the uranium, since to face the main constraints low inherent P in soil and plants, rock phosphates are likely to be more extensively disseminated in the agriculture and these phosphorus sources carry uranium from their structures to the environment and human food chain as well [21].

Uranium is the heaviest natural element in the nature; it is hazardous element in man and animal health, not just it presents radioactivity but also it presents metallotoxicity once it is a heavy metal. Furthermore, uranium presents many radionuclides with high radioactivity and energy [22].

Health implications of human exposure to uranium are well documented: cancer, liver and kidney diseases and reproduction impairment [21,22].

Uranium in nature is more plentiful than silver (Ag) and about as abundant as arsenic (As). It is found in very small amounts in the form of minerals, especially in rocks, soil, water, air, plants and animal tissues that could be consumed as food containing varying amounts of uranium [23].

Practically every rock phosphate contains uranium in its structure [16,24]. The amounts of this and others hazardous substances vary widely among phosphates sources and it may vary even in the same deposit. Thus, mining, milling, industrializing and using phosphate products in soil and animal nutrition are anthropogenic activities increasing the potential for human exposure to uranium [24,25].

German studies in the 1970's pointed-out the evidence for a raising uranium presence in rivers and groundwater in regions with intensive use of phosphate fertilizers in agriculture (figure 1). Uranium derived from phosphate fertilizers is likely to be adsorbed on the uppermost soil layers and its content on the water is correlated to the HCO_3^- content in the river [25].

Uranium content can be determined by the nuclear method of the Neutron Activation Analysis. This is a precise, fast (short turn-around), sensitive and non-destructive method [2].

4. Material and procedures

Phosphate sample preparation: phosphates were acquired in the local market of Minas Gerais. Aliquots of 100 grams were randomly taken for each product pack to be grounded to obtain a particle size of 200 Tyler mesh (75 μm) establishing similar conditions for all samples (99% of conformity of the particle size of each product). Aliquot of 1000 mg of each grounded product was weighted and sealed in small polystyrene capsules.

Animal breeding: Two groups of twelve (6 females and 6 males) New Zealand white rabbits (30 days of life) were selected and separated in two groups housed individually receiving a different phosphorus source (dicalcium phosphate and bovine bone meal).

Two rabbit feeds were designed to allow the introduction of the P source and to offer sufficient nutrient intake to meet rabbit nutritive requirements. Each one of the feeds had the same 98 percent (dry basis) of fiber, energy, and amino acids. Both formulations were based in raw materials as %, dry basis: Alfalfa meal 34.63, Soy oil 1.00, Sugar cane 2.00, Salt 0.50, Lysine 0.25, Methionine 0.04, Limestone 1.0, Premix 0.40, Maize 6.05, Wheat straw 25.0, Soybean meal 12.13, Maize by products 15.00 and the remaining 2.0 from the selected P source – dicalcium phosphate and bone meal, both materials analyzed in first part of the experiment.

Feeds were processed in order to turn the mixed products into a compact mixture. After that, the meal was conditioned by mixing it on dry steam in a conditioner; this conditioned product was pressed by rolls to pass through the holes of the pelleting die which shapes the meal into the final pellet shape of 3.00 mm to permit a good balance between pellet quality and good intestinal motility.

Animals were housed in stainless steel cages with a fenestrated floor to allow feces to drop through into a pan. Absorbent material was placed in the pan to collect urine and minimize ammonia release due to the bacterial breakdown of urea.

Good quality water was provided through a nipple-drinking system that provides water at all times. Food was provided by a J-hopper attached to the front of the cage. J-hopper prevents the rabbit from defecating in their food. Animals were fed *ad libitum* from 30 to 72 days - age considered as ideal slaughtering that allows rabbit to reach its commercial live weight, i.e. 2.0 kilograms.

The 24 rabbits were slaughtered via humanitarian euthanasia and their *longissimus dorsi* muscles were extracted and prepared by freeze-dry process in order to be irradiated. Approximately 100 g of each specimen was frozen at –70°C and lyophilized. Each freeze-dried sample was powdered and homogenized and around 1000 mg was taken into polyethylene irradiation vials.

Irradiation: Each one of the full-filled capsules with samples was placed in a polyethylene container for the pneumatic transporting system. Individually, samples and standards in the vials were transported into the neutron flux using the pneumatic transport system of the reactor IPR-R1 in the CDTN/CNEN (Centre of the Nuclear Technology Development) in Belo Horizonte, Brazil. The reactor was operated at 100 kW-thermal power under a neutron flux of 6.6×10^{11} neutrons.cm^2.s^{-1}. Additionally, phosphates were studied by the well-established Colorimetric Method to assess their phosphorus - P_2O_5 content, at the EC-4 Sector of the Nuclear Technology Development Centre, institute from the Brazilian Nuclear Energy Commission (CDTN/CNEN).

5. Analytical technique applied on elemental determination

The neutron activation technique (NA) is based on nuclear properties of the nucleus of the atom, radio- activity, and the interaction of radiation with matter. The simplest description of the technique says that when one natural element is submitted to a neu- tron flux, the reaction (n,γ) occurs. The radionuclide formed emits gamma radiation, which can be meas- ured by suitable equipment. About 70% of the ele- ments have nuclides possessing properties suitable for neutron activation analysis. At the Nuclear Technology Development Centre (CDTN), there is a nuclear reac- tor TRIGA MARK I IPR-R1 that allows the application of this technique [7].

The k0-instrumental neutron activation analysis (k0-INAA,) a variation of NA in which the sample is irradiated without previous chemical preparation was used in this study. This specific method is based on nuclear constants—the k0 factors and some reactor parameters.

Rabbit tissues were irradiated in the reactor TRIGA MARK I IPR-R1. At 150 kW the thermal neutron flux is 6.6×10^{11} neutrons cm^2 s^{-1}. The samples were irradiated simultaneously with standards of gold and sodium as comparators, and the reference materials. The elements were determined through three schemes of irradiation: 5 minutes to detect the short half-life radionuclides; 4 hours to detect the medium, and 20 hours, the long half-life radionuclides.

After suitable decay time, the gamma spectroscopy was performed in a HPGe detector, 10% of efficiency, FWHM 1.85 keV and ^{60}Co, 1332 keV, connected to a multichannel analyzer. The calculations were based on the reactor parameters: k0 constants using the Solcoy Sofware ©.

6. Results

Phosphate	Total U [μg.g⁻¹]	Ratio P:U
Amm. polyphosphate, 45% P_2O_5, *Brazil*	37 ± 4[a]	5580:1
Super-simple phosphate, 17% P_2O_5, *Brazil*	49 ± 5[a]	1485:1
Dicalcium phosphate, 45%, P_2O_5, *Brazil*	187 ± 9[a]	1050:1
Monoamm. phosphate, 51% P_2O_5, *Brazil*	183 ± 9[a]	1215:1
Super-triple phosphate, 45% P_2O_5, *Brazil*	34 ± 4[a]	3975:1
Israeli rock phosphate, 31% P_2O_5, *Israel*	145 ± 7[a]	935:1
Rock phosphate, 31% P_2O_5, *Florida, USA*	59[b]	2300:1
Rock phosphate, 28% P_2O_5, *Tanzania*	390[b]	303:1
Rock phosphate, 29% P_2O_5, *Mali*	123[b]	1030:1
Bovine Bone Meal	1.0 ± 0.8[a]	27650:1

[a] U experimental results by Delayed Neutrons Technique, P by Colorimetric Method

[b] U and P data extracted from FAO/IAEA [16]

Table 5. U content and the ratio [phosphorus: uranium] in the tested phosphates

Phosphorus Source	U Concentration [μg.g⁻¹]
Dicalcium Phosphate	1.25 ± 0.45[a]
Bovine Bone Meal	0.91 ± 0.29[a]

[a] The results were evaluated ($p \leq 0.05$) by t-test using Microsoft Excel 2000 software [26]. Means with the same letter ([a]) are not significantly different.

Table 6. Uranium content in the *longissimus dorsi* muscle from rabbits fed with two different sources (as 2% in the feed) of phosphorus, n=12. The data are presented as mean ± SD, raw basis.

7. Discussion

Concentrations of uranium are quite variable in the phosphate products. The average ratio of phosphorus to uranium in phosphates appeared not to be only related to their origin since all tested Brazilian phosphates are from igneous deposits (ratio varying from 1000-4000 atoms of phosphorus to 1 atom of uranium), instead the foreign data are related exclusively to sedimentary rocks (ratio varying from 300-2400 atoms of phosphorus to 1 atom of Uranium). Differences amongst tested products are the region of exploitation of phosphate ores that implies different ages of mineralization, deposit types and accessory minerals associated that may vary in phosphates of the same origin, and finally they are separated by the routes of production of each final product for those are industrialized.

No significant difference was observed in uranium presence in the muscle tissues from rabbits (tab. 6) receiving dicalcium phosphate and those animals receiving bovine bone meal: average uranium content are 1.25 ± 0.45 µg.g^{-1} for dicalcium phosphate fed group and 0.91 ± 0.29 µg.g^{-1} for the bovine bone meal fed group. Possibly, *longissimus dorsi* muscle is not a target tissue for uranium in mammals. Indeed, uranium is often associated to replace calcium in the bone hydroxyapatites; both are considered mutual inorganic cation exchangers in apatite [21].

8. Conclusions

In conclusion, Nuclear analytical techniques have been helping to improve the human welfare in terms of health acting as outstanding tools working around the triade: diagnosis, prevention and treatment. In many analyses, Nuclear techniques are not merely an option, but are the only feasible solution to address some assessments.

Following the '*Precautionary Principle*' [27] what states that if an action might theoretically or logically cause harm, then those who wish to undertake this action (of using these substances in this case) have a greater moral responsibility to demonstrate good evidence that the use does not cause any harm, not now not in the long run.

Nuclear Analytical Techniques have a long and useful history in Life Sciences. These applications in Agriculture are a warranty for mutual benefits for both arenas, as long Nuclear Power generation is currently in a moratorium after Fukushima-Daichi 2011 disaster, and Agriculture for food and energy production is boosting and evolving.

Disclaimer

Data and statements expressed in this paper are those from the authors, and do not necessarily reflect organisations with which the authors are affiliated. The authors do not endorse any equipment or material cited herein.

Acronyms and abbreviations

AOAC Association of Official Analytical Chemists

ASTDR Agency for Toxic Substances and Disease Registry

CERCLA Comprehensive Environmental Response, Compensation, and Liability Act

CDTN *Centro de Desenvolvimento da Tecnologia Nuclear*, Nuclear Technology Development Centre, Belo Horizonte, Brazil

CNEN *Comissão Nacional de Energia Nuclear* Nuclear Energy (Brazilian) National Commission

CRM Certified Reference Material

DNPM *Departamento Nacional de Produção Mineral*, National Department for Mineral Production, Brazil

IAEA International Atomic Energy Agency

IPEN *Instituto de Pesquisas Energéticas Nucleares* Nuclear Energy Research Institute, São Paulo, Brazil

USDA United States Department of Agriculture

USEPA United State Environment Protection Agency

USP University of Sao Paulo

pg picogram

pmol picomole

PHS Public Health Service

PMR proportionate mortality ratio

ppb parts per billion

ppm parts per million

ppt parts per trillion

yr year

> greater than

= equal to

< less than

% percent

α alpha

β beta γ gamma

δ delta

μm micrometer

μg microgram

Acknowledgements

This project is partially supported by: the Minas Gerais State Research Funding Agency FAPEMIG, the Brazilian Ministry of Science and Technology of Brazil via CNPq and the Brazilian Ministry of Education via CAPES/School of Veterinary from the Universidade Federal de Minas Gerais. We appreciate every support from CDTN/CNEN in particular we wish to thank its employee, Dr. Maria A.B.C. Menezes who supported us in our analyses.

Author details

A.C. Avelar*, W.M. Ferreira and M.A.B.C. Menezes

*Address all correspondence to: avelara@ufmg.br

DZOO Department of Animal Sciences, Universidade Federal de Minas Gerais Avenida Antonio Carlos, Campus UFMG, Pampulha, Belo Horizonte, Brazil

References

[1] Atwater, W. O, & Woods, C. D. American Food Materials, Washington: Government Printing Office, (1896). pp.

[2] Parry, S. J. Handbook of neutron activation analysis. Woking: Viridian Publishing, (2003). p.

[3] Bode, P. Opportunities for innovation in neutron activation analysis Peter, Journal of Radioanalytical and Nuclear Chemistry, Published on line: 03 June (2011). pp.

[4] ManahanStanley E. Fundamentals of Environmental Chemistry Boca Raton: CRC Press LLC,(2001). pp.

[5] Avelar, A. C. Nuclear engineering education in Brazil: Review and prospects DOI:s10967-007-7282-8 Journal of Radioanalytical and Nuclear Chemistry, (2009). , 279(1)

[6] Fávaro, D. I. T, Afonso, C, Vasconcellos, M. B. A, & Cozzolino, S. M. F. Determinação de elementos minerais e traços por ativação neutrônica, em refeições servidas no restaurante da Faculdade de Saúde Pública/USP (in Portughese), Ciência e Tecnologia de Alimentos, n. 1, (2000). , 20, 176-182.

[7] Avelar, A. C, Veado, J. C. C, & Menezes, M. A. B. C. Study of Essential Elements in Cattle Tissues from a tropical country using Neutron Activation Analysis Food And Nutrition Bulletin, (2002). , 23, 237-240.

[8] Marçal, W. S, Gomes, G. P, Nascimento, M. R. L, & Moreno, A. M. Evaluation of Phosphorus Sources for Cattle Feed Supplementation. Arq. Inst. Biol., São Paulo, n.3, (2003). , 70, 255-258.

[9] Spiegl, C. J. (1949). Pharmacology and toxicology of uranium compounds. McGraw-Hill Book Company, Inc. , 532-547.

[10] Dygert, H. P. (1949). Pharmacology and toxicology of uranium compounds. New York, NY: McGraw-Hill Book Inc., , 603-613.

[11] Tokarnia, C. H, Dobereiner, J, & Peixoto, P. V. (2000). Deficiências minerais em animais de fazenda, principalmente bovinos, Pesq. Vet. Bras., , 20(3), 127-138.

[12] Sanchez, P. A. (2002). Soil fertility and hunger in Africa. Science, , 295, 2019-2020.

[13] Sauchelli, V. (1962). Origin of Processing of Phosphate Rock with Particular Reference to Benefication, The International Fertiliser Society : York (UK) 30pp.

[14] Guimarães, R. C, Araújo, A. C, & Peres, A. E. C. (2005). Reagents in igneous phosphate ores flotation Minerals Engineering, , 18, 199-204.

[15] Huxley, A. *Brave New World*, ((1932). Available at http://www.idph.net,accessed in June 11, 2012, 176 pp.

[16] FAO/IAEA- Food and Agriculture Organization and International Atomic Energy Agency ((2004). Use of phosphate rocks for sustainable agricultureFAO- Food and Agriculture Organization, 1st ed., FAO: Roma, 148pp.

[17] DNPM National Department of Mineral Production ((2010). Mineral Year Book (in Portuguese), DNPM, Brazilian Ministry of Mines and Energy, Brasilia, 871pp.

[18] Sindirações *Informative Bulletin of the Animal Feed Sector*in Portuguese), Boletim Informativo do Setor de Alimentação Animal, 8Available at: http://sindiracoes.org.br/wp-content/uploads/2012/05/sindiracoes_boletim-informativo-versao-portugues-atual-maio2012.pdfAccessed in 09/13/2012, 2012.

[19] ATSDR Agency for Toxic Substances and Disease Registry ((2012). in Portughese) *2011 Priority List of Hazardous Substance* 19 ppConsulted in 09/06/2012, Available at: http://http://www.atsdr.cdc.gov/SPL/resources/ ATSDR_2011_SPL_Detailed_Data_Table.pdf

[20] ATSDR Agency for Toxic Substances and Disease Registry ((2012). *Support Document to the 2011 Priority List of Hazardous Substances that will be the Subject of Toxicological Profiles*Atlanta. 9pp.

[21] ATSDR Agency for Toxic Substances and Disease Registry of United States ((1999). *Toxicological profile for uranium*Atlanta, USA: U.S. Department of Health and Human Services, 432 p.

[22] Magdo, H. S, Forman, J, Graber, N, Newman, B, Klein, K, Satlin, L, Amler, R. W, Winston, J, & Landrigan, J. P. (2007). Grand Rounds: Nephrotoxicity in a young child exposed to uranium from contaminated well water, Environ. Health Perspect., , 115(8), 1237-1241.

[23] Shiraishi, K, & Yamamoto, M. (1995). Dietary ^{232}Th and ^{238}U intakes for Japanese as obtained in a market basket study and contributions of imported foods to internal doses. J.Radioanal. Nucl. Chem. , 196(1), 89-96.

[24] Scholten, L. C, & Timmermans, C. W. M. (1996). Natural radioactivity in phosphate fertilizers Fertilizer Research , 43, 103-107.

[25] Mangini, A, Sonntag, C, Bertsch, G, & Müller, E. (1979). Evidence for a higher natural uranium content in world rivers, Nature, , 278, 337-339.

[26] Miller, J. N, & Miller, J. C. (2010). Statistics and Chemometrics for Analytical Chemistry, 6th Edition, Essex, 278 pp.

[27] Mills, D. S. Ed. ((2010). The Encyclopedia of Applied Animal Behaviour and Welfare, CABI, Oxfordshire, 1081 pp.

Chapter 7

Concepts, Instrumentation and Techniques of Neutron Activation Analysis

Lylia Hamidatou, Hocine Slamene, Tarik Akhal and Boussaad Zouranen

Additional information is available at the end of the chapter

1. Introduction

Analytical science to develop the methodology for the investigation of properties and structure of matter at level of single nucleus, atom and molecule, and scientific analysis to determine either chemical composition or elemental contents in a sample are indispensable in basic research and development, as well as in industrial applications.

Following the discovery of neutron by J. Chadwick in 1932 (Nobel prize, 1935) and the results of F. Joliot and I. Curie in 1934, neutron activation analysis was first developed by G. Hevesy and H. Levi in 1936. They used a neutron source (226Ra + Be) and a radiation detector (ionization chamber) and promptly recognized that the element Dy (dysprosium) in the sample became highly radioactive after exposure to the neutron source. They showed that the nuclear reaction may be used to determine the elements present in unknown samples by measuring the induced radioactivity.

Thereafter, the development of the nuclear reactors in the 1940s, the application of radiochemical techniques using low resolution scintillation detectors like NaI (Tl) in the 1950s, the development of semiconductor detectors (Ge, Si, etc.) and multichannel analyzer in the 1960s, and the advent of computers and relevant software in the 1970s, the nuclear technique has advanced to become an important analytical tool for determination of many elements at trace level. In spite of the developments in other chemical techniques, the simplicity and selectivity, the speed of operation, the sensitivity and accuracy of NAA have become and maintained its role as a powerful analytical technique. In 2011, Peter Bode describes in his paper "Neutron activation analysis: A primary method of measurements", the history of the development of NAA overall the world [1].

Nowadays, there are many elemental analysis methods that use chemical, physical and nuclear characteristics. However, a particular method may be favoured for a specific task, depending on the purpose. Neutron activation analysis (NAA) is very useful as sensitive analytical technique for performing both qualitative and quantitative multielemental analysis of major, minor and traces components in variety of terrestrial samples and extra-terrestrial materials. In addition, because of its accuracy and reliability, NAA is generally recognized as the "referee method" of choice when new procedures are being developed or when other methods yield results that do not agree. It is usually used as an important reference for other analysis methods. Worldwide application of NAA is so widespread it is estimated that approximately 100,000 samples undergo analysis each year.

The method is based on conversion of stable atomic nuclei into radioactive nuclei by irradiation with neutrons and subsequent detection of the radiation emitted by the radioactive nuclei and its identification. The basic essentials required to carry out an analysis of samples by NAA are a source of neutrons, instrumentation suitable for detecting gamma rays, and a detailed knowledge of the reactions that occur when neutrons interact with target nuclei. Brief descriptions of the NAA method, reactor neutron sources, and gamma-ray detection are given below.

This chapter describes in the first part the basic essentials of the neutron activation analysis such as the principles of the NAA method with reference to neutron induced reactions, neutron capture cross-sections, production and decay of radioactive isotopes, and nuclear decay and the detection of radiation. In the second part we illustrated the equipment requirements neutron sources followed by a brief description of Es-Salam research reactor, gamma-ray detectors, and multi-channel analysers. In addition, the preparation of samples for neutron irradiation, the instrumental neutron activation analysis techniques, calculations, and systematic errors are given below. Some schemes of irradiation facilities, equipment and materials are given as examples in this section.

Finally, a great attention will be directed towards the most recent applications of the INAA and k0-NAA techniques applied in our laboratory. Examples of such samples, within a selected group of disciplines are milk, milk formulae and salt (nutrition), human hair and medicinal seeds (biomedicine), cigarette tobacco (environmental and health related fields) and iron ores (exploration and mining).

All steps of work were performed using NAA facilities while starting with the preparation of samples in the laboratory. The activation of samples depends of neutron fluence rate in irradiation channels of the Algerian Es-Salam research reactor. The radioactivity induced is measured by gamma spectrometers consist of germanium based semiconductor detectors connected to a computer used as a multichannel analyser for spectra evaluation and calculation. Sustainable developments of advanced equipment, facilities and manpower have been implemented to establish a state of the art measurement capability, to implement several applications, etc.

2. Neutron activation analysis

Neutron activation analysis (NAA) is a nuclear process used for determining the concentrations of elements in a vast amount of materials. NAA relies on excitation by neutrons so that the treated sample emits gamma-rays. It allows the precise identification and quantification of the elements, above all of the trace elements in the sample. NAA has applications in chemistry but also in other research fields, such as geology, archaeology, medicine, environmental monitoring and even in the forensic science.

2.1. Basis principles

The sequence of events occurring during the most common type of nuclear reaction used for NAA, namely the neutron capture or (n, gamma) reaction, is illustrated in Figure 1. Creation of a compound nucleus forms in an excited state when a neutron interacts with the target nucleus via a non-elastic collision. The excitation energy of the compound nucleus is due to the binding energy of the neutron with the nucleus. The compound nucleus will almost instantaneously de-excite into a more stable configuration through emission of one or more characteristic prompt gamma rays. In many cases, this new configuration yields a radioactive nucleus which also de-excites (or decays) by emission of one or more characteristic delayed gamma rays, but at a much lower rate according to the unique half-life of the radioactive nucleus. Depending upon the particular radioactive species, half-lives can range from fractions of a second to several years.

In principle, therefore, with respect to the time of measurement, NAA falls into two categories: (1) prompt gamma-ray neutron activation analysis (PGNAA), where measurements take place during irradiation, or (2) delayed gamma-ray neutron activation analysis (DGNAA), where the measurements follow radioactive decay. The latter operational mode is more common; thus, when one mentions NAA it is generally assumed that measurement of the delayed gamma rays is intended. About 70% of the elements have properties suitable for measurement by NAA.

The PGAA technique is generally performed by using a beam of neutrons extracted through a reactor beam port. Fluxes on samples irradiated in beams are in the order of one million times lower than on samples inside a reactor but detectors can be placed very close to the sample compensating for much of the loss in sensitivity due to flux. The PGAA technique is most applicable to elements with extremely high neutron capture cross-sections (B, Cd, Sm, and Gd); elements which decay too rapidly to be measured by DGAA; elements that produce only stable isotopes (e.g. light elements); or elements with weak decay gamma-ray intensities. 2D, 3D-analysis of (main) elements distribution in the samples can be performed by PGAA.

DGNAA (sometimes called conventional NAA) is useful for the vast majority of elements that produce radioactive nuclides. The technique is flexible with respect to time such that the sensitivity for a long-lived radionuclide that suffers from interference by a shorter-lived radionuclide can be improved by waiting for the short-lived radionuclide to decay or quite the contrary, the sensitivity for short-lived isotopes can be improved by reducing the time

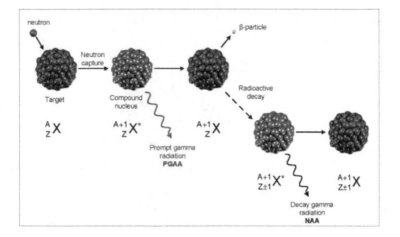

Figure 1. Diagram illustrating the process of neutron capture by a target nucleus followed by the emission of gamma rays.

irradiation to minimize the interference of long-lived isotopes. This selectivity is a key advantage of DGNAA over other analytical methods.

In most cases, the radioactive isotopes decay and emit beta particles accompanied by gamma quanta of characteristic energies, and the radiation can be used both to identify and accurately quantify the elements of the sample. Subsequent to irradiation, the samples can be measured instrumentally by a high resolution semiconductor detector, or for better sensitivity, chemical separations can also be applied to reduce interferences. The qualitative characteristics are: the energy of the emitted gamma quanta ($E\gamma$) and the half life of the nuclide ($T_{1/2}$). The quantitative characteristic is: the $I\gamma$ intensity, which is the number of gamma quanta of energy $E\gamma$ measured per unit time.

The n-gamma reaction is the fundamental reaction for neutron activation analysis. For example, consider the following reaction:

^{58}Fe $+^1$ n $\rightarrow ^{59}$ Fe + Beta$^-$+ gamma rays

^{58}Fe is a stable isotope of iron while ^{59}Fe is a radioactive isotope. The gamma rays emitted during the decay of the ^{59}Fe nucleus have energies of 142.4, 1099.2, and 1291.6 KeV, and these gamma ray energies are characteristic for this nuclide (see figure 2) [2]. The probability of a neutron interacting with a nucleus is a function of the neutron energy. This probability is referred to as the capture cross-section, and each nuclide has its own neutron energy-capture cross-section relationship. For many nuclides, the capture cross-section is greatest for low energy neutrons (referred to as thermal neutrons). Some nuclides have greater capture cross-sections for higher energy neutrons (epithermal neutrons). For routine neutron activation analysis we are generally looking at nuclides that are activated by thermal neutrons.

The most common reaction occurring in NAA is the (n,γ) reaction, but also reactions such as (n,p), (n,α), (n,n') and $(n,2n)$ are important. The neutron cross section, σ, is a measure for the

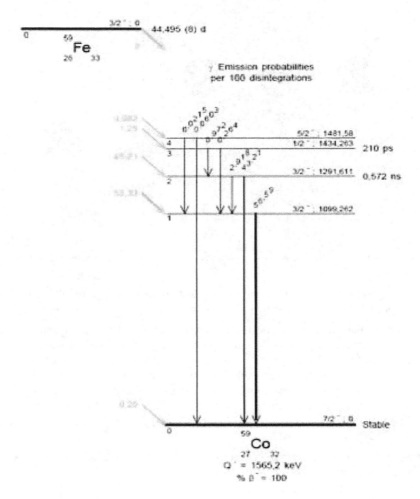

Figure 2. Decay scheme of ^{59}Fe.

probability that a reaction will take place, and can be strongly different for different reaction types, elements and energy distributions of the bombarding neutrons. Some nuclei, like 235U are fissionable by neutron capture and the reaction is denoted as (n,f), yielding fission products and fast (highly energetic) neutrons [1].

Neutrons are produced via

- Isotopic neutron sources, like ^{226}Ra(Be), ^{124}Sb(Be), ^{241}Am(Be), ^{252}Cf. The neutrons have different energy distributions with a maximum in the order of 3–4 MeV; the total output is typically 10^5–10^7 s^{-1} GBq^{-1} or, for ^{252}Cf, 2.2 10^{12} s^{-1}g^{-1}.

- Particle accelerators or neutron generators. The most common types are based on the acceleration of deuterium ions towards a target containing either deuterium or tritium, resulting in the reactions 2H(2H,n)3He and 3H(2H,n)4He, respectively. The first reaction, often denoted as (D,D), yields monoenergetic neutrons of 2.5 MeV and typical outputs in the order of 10^8–10^{10} s^{-1}; the second reaction (D,T) results in monoenergetic neutrons of 14.7 MeV and outputs of 10^9–10^{11} s^{-1}.

- Nuclear research reactors. The neutron energy distribution depends on design of the reactor and its irradiation facilities. An example of an energy distribution in a light water moderated reactor is given in Fig. 2.3 from which it can be seen that the major part of the neutrons has a much lower energy distribution that in isotopic sources and neutron generators. The neutron output of research reactors is often quoted as neutron fluence rate in an irradiation facility and varies, depending on reactor design and reactor power, between 10^{15} and 10^{18} m^{-2} s^{-1}.

Owing to the high neutron flux, experimental nuclear reactors operating in the maximum thermal power region of 100 kW -10 MW with a maximum thermal neutron flux of 10^{12}-10^{14} neutrons cm^{-2} s^{-1} are the most efficient neutron sources for high sensitivity activation analysis induced by epithermal and thermal neutrons. The reason for the high sensitivity is that the cross section of neutron activation is high in the thermal region for the majority of the elements. There is a wide distribution of neutron energy in a reactor and, therefore, interfering reactions must be considered. In order to take these reactions into account, the neutron spectrum in the channels of irradiation should be known exactly. E.g. if thermal neutron irradiations are required, the most thermalized channels should be chosen.

Although there are several types of neutron sources (reactors, accelerators, and radioisotopic neutron emitters) one can use for NAA, nuclear reactors with their high fluxes of neutrons from uranium fission offer the highest available sensitivities for most elements. Different types of reactors and different positions within a reactor can vary considerably with regard to their neutron energy distributions and fluxes due to the materials used to moderate (or reduce the energies of) the primary fission neutrons. This is further elaborated in the title "Derivation of the measurement equation". In our case, the NAA method is based on the use of neutron flux in several irradiation channels of Es-Salam Research reactor. In 2011, Hamidatou L et Al., reported "Experimental and MCNP calculations of neutron flux parameters in irradiation channel at Es-Salam reactor" the core modelling to calculate neutron spectra using experimental and MCNP approaches. The Es-Salam reactor was designed for a thermal power output of 15 Mw, with 72 cylindrical cluster fuel elements; each fuel element consists of 12 cylindrical rods of low enriched UO2. In addition the both of fuel throttle tube of the cluster and fuel element tube encloses heavy water as moderator and coolant. The fuel elements are arranged on a heavy water square lattice. The core of the reactor is constituted by a grid containing 72 fuel elements, 12 rods for reactivity control and two experimental channels.

There is also a heavy water in the middle of the core including five experimental channels called inner reflector, In addition, all fuel elements have a reflector at each end called upper and lower reflector. The core is reflected laterally by heavy water maintained in aluminium tank followed by the graphite.

2.2. Neutron activation analysis procedure

In the majority of INAA procedures thermal reactor neutrons are used for the activation: neutrons in thermal equilibrium with their environment. Sometimes activation with epithermal reactor neutrons (neutrons in the process of slowing down after their formation from fission of 235U) is preferred to enhance the activation of elements with a high ratio of resonance neutron cross section over thermal neutron cross section relatively to the activation of elements with a lower such a ratio. In principle materials can be activated in any physical state, viz. solid, liquid or gaseous. There is no fundamental necessity to convert solid material into a solution prior to activation; INAA is essentially considered to be a non-destructive method although under certain conditions some material damage may occur due to thermal heating, radiolysis and radiation tracks by e.g. fission fragments and α-radiation emitting nuclei. It is essential to have more than two or three qualified full-time member of the staff with responsibility for the NAA facilities. They should be able to control the counting equipment and have good knowledge of basic principles of the technique. In addition, the facility users and the operators must establish a good channel of communication. Other support staff will be required to maintain and improve the equipment and facility. It seems, therefore, a multi-disciplinary team could run the NAA system well.

The analytical procedure is based on four steps:

Step 1: sample preparation (Figure 3) means in most cases only heating or freeze drying, crushing or pulverization, fractionating or pelletizing, evaporation or pre-concentration, put through a sieve, homogenising, weighing, washing, check of impurities (blank test), encapsulation and sealing irradiation vial, as well as the selection of the best analytical process and the preparation of the standards. The laboratory ambiance is also important for preservation and storage of the samples. Standardization is the basis for good accuracy of analytical tools and often depends on particular technology, facility and personnel. For production of accurate data, careful attention to all possible errors in preparing single or multi-element standards is important, and standards must be well chosen depending on the nature of the samples.

Step 2: irradiation of samples can be taken from the various types of neutron sources according to need and availability. For the INAA, one pneumatic transfer system installed in the horizontal channel at Es-Salam research reactor for short irradiation of samples (Figure 4). In addition, two vertical channels located in different sites of the heavy water moderator and the graphite reflector have been used for long irradiations. The neutron spectrum parameters at different irradiation channels such as alpha, f, Tn, etc are experimentally determined using cadmium ratio, cadmium cover, bare triple monitor and bi-isotopic methods using HΦgdhal convention and Westcott formalism Table 1 and Table 2. The calibration of the irradiation positions has been carried out to implement the k_0-NAA in our laboratory.

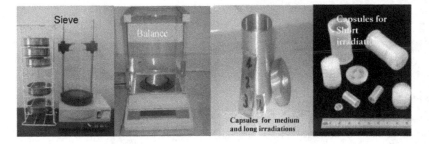

Figure 3. Some instruments and materials used for the sample preparation.

method	α	f	$r(a)\sqrt{T_n/T_0}$
Cd-ratio	0.026±0.012	28.4±1.6	0.038±0.004
Cd-covered	0.024±0.010	28.7±2.1	-
Bare triple monitor	0.030±0.008	28.6±1.8	-
Bare bi-isotopic	-	29.5±2.5	0.036±0.003
Average	0.027±0.010	28.8±2.0	0.037±0.003

Table 1. The parameters α, f and $r(a)\sqrt{T_n/T_0}$ obtained by different methods.

parameter	α	f	Tn(°C)	Rcd(Au)	$r(a)\sqrt{T_n/T_0}$
Measured value	0.027±0.010	28.8±2.0	34±1.8	2.93±0.32	0.037±0.003

Table 2. Neutron spectrum parameters in the irradiation site at es-Salam research reactor.

Step 3: after the irradiation the measurement is performed after a suitable cooling time (t_c). In NAA, nearly exclusively the (energy of the) gamma radiation is measured because of its higher penetrating power of this type of radiation, and the selectivity that can be obtained from distinct energies of the photons - differently from beta radiation which is a continuous energy distribution. The interaction of gamma- and X-radiation with matter results, among others, in ionization processes and subsequent generation of electrical signals (currents) that can be detected and recorded.

The instrumentation used to measure gamma rays from radioactive samples generally consists of a semiconductor detector, associated electronics, and a computer-based multi-channel analyzer (MCA/computer).

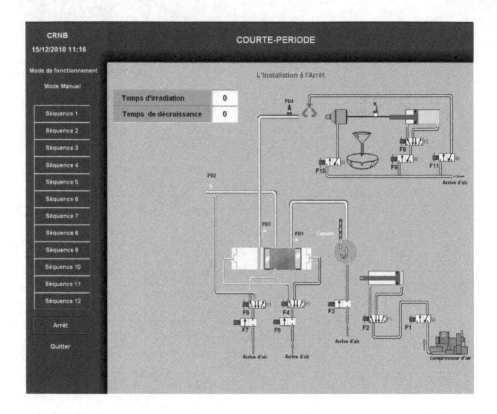

Figure 4. Pneumatic system for short irradiations using a thermal neutron flux at Es-Salam research reactor.

Most NAA labs operate one or more hyper-pure germanium (HPGe) detectors, which operate at liquid nitrogen temperature (77 K). Although HPGe detectors come in many different shapes and sizes, the most common shape is coaxial. These detectors are very useful for measurement of gamma rays with energies in the range from about 60 keV to 3.0 MeV. The two most important characteristics a HPGe detector are its resolution and efficiency. Other characteristics to consider are peak shape, peak-to-Compton ratio, pulse rise time, crystal dimensions or shape, and price. The detector's resolution is a measure of its ability to separate closely spaced peaks in the spectrum, and, in general, the resolution is specified in terms of the full width at half maximum (FWHM) of the 122 keV photopeak of ^{57}Co and the 1,332 keV photopeak of ^{60}Co. For most NAA applications, a detector with 0.5 keV resolution or less at 122 keV and 1.8 keV or less at 1,332 keV is sufficient. Detector efficiency for a given detector depends on gamma-ray energy and the sample and detector geometry, i.e. subtended solid angle. Of course, a larger volume detector will have a higher efficiency.

At Es-Salam NAA Lab, four gamma-ray spectrometers of Canberra for which one of them consists of a HPGe detector 35% relative efficiency connected with Genie 2k Inspector and the three other spectrometers are composed of detectors (30, 35 and 45 % relative efficiency) connected with a three Lynx® Digital Signal Analyser, It is a 32K channel integrated signal analyzer based on advanced digital signal processing (DSP) techniques. All spectrometers operate with Genie™2000 spectroscopy software. A radiation detector therefore consists of an absorbing material in which at least part of the radiation energy is converted into detectable products, and a system for the detection of these products. Figure 5 illustrates Gamma-ray spectroscopy systems. The detectors are kept at liquid nitrogen temperatures (dewers under cave). The boxes in the left and in the right of the computer are the Lynx Digital Spectrometer Processing.

Figure 5. Gamma-ray spectroscopy systems in NAA/CRNB laboratory.

Step 5: Measurement, evaluation and calculation involve taking the gamma spectra and the calculating trace element concentrations of the sample and preparation of the NAA report.

In this part of work, Peter bode describes clearly in his paper [1] the analysis procedure of gamma-spectrum to the determination of the amount of element in sample. The acquisition of gamma spectrum Fig.6 and Fig.7 via the spectroscopy system Fig. 5 is analyzed to identify the radionuclides produced and their amounts of radioactivity in order to derive the target elements from which they have been produced and their masses in the activated sample. The spectrum analysis starts with the determination of the location of the (centroids of the) peaks. Secondly, the peaks are fitted to obtain their precise positions and net peak areas. The Analytical protocol adopted in our NAA laboratory is presented in Fig.8.

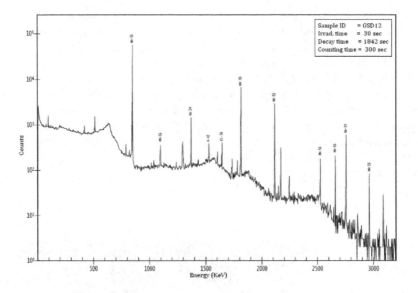

Figure 6. Gamma-ray spectrum showing several short-lived elements measured in a CRM-DSD-12 standard irradiated at Es-salam research reactor for 30 seconds, decayed for 30.7 minutes, and counted for 5 minutes with an HPGe detector.

The positions – often expressed as channel numbers of the memory of a multi-channel pulse height analyzer – can be converted into the energies of the radiation emitted; this is the basis for the identification of the radioactive nuclei. On basis of knowledge of possible nuclear reactions upon neutron activation, the (stable) element composition is derived. The values of the net peak areas can be used to calculate the amounts of radioactivity of the radionuclides using the full energy photopeak efficiency of the detector.

The amounts (mass) of the elements may then be determined if the neutron fluence rate and cross sections are known. In the practice, however, the masses of the elements are determined from the net peak areas by comparison with the induced radioactivity of the same neutron activation produced radionuclides from known amounts of the element of interest. The combination of energy of emitted radiation, relative intensities if photons of different energies are emitted and the half life of the radionuclide is unique for each radionuclide, and forms the basis of the qualitative information in NAA. The amount of the radiation is directly proportional to the number of radioactive nuclei produced (and decaying), and thus with the number of nuclei of the stable isotope that underwent the nuclear reaction. It provides the quantitative information in NAA.

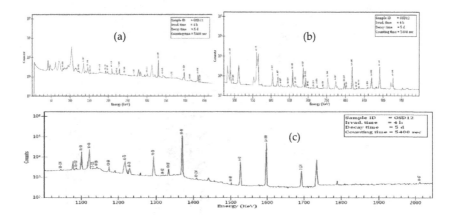

Figure 7. Gamma-ray spectrum (**a**) from 0 to 450 keV, (**b**) from 450 to 1000 keV and (**c**) from 1000 to 2000 keV: showing medium- and long-lived elements measured in a sample of CRM-GSD-12 standard irradiated at Es-salam research reactor for 4 hours, decayed for 5 days, and counted for 90 minutes on a HPGe detector.

The measured in NAA – the quantity intended to be measured – is the total mass of a given element in a test portion of a sample of a given matrix in all physico-chemical states. The quantity 'subject to measurement' is the number of disintegrating nuclei of a radionuclide. The measurement results in the number of counts in a given period of time, from which the disintegration rate and the number of disintegrating nuclei is calculated; the latter number is directly proportional to the number of nuclei of the stable isotope subject to the nuclear reaction, and thus to the number of nuclei of the element, which finally provides information on the mass and amount of substance of that element (see Eq. 16). An example of typical ranges of experimental conditions is given in Table 3 [1].

In practice, our laboratory proceeds in the treatment of spectra and calculation of elemental concentrations of analyzed samples according the approach illustrated in figure 8.

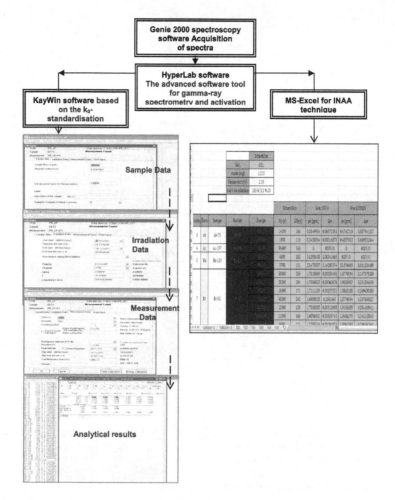

Figure 8. Analytical protocol adopted in NAA/CRNB laboratory [13].

Test portion mass : 5-500 mg			
Neutron fluence rates available $10^{16} - 10^{18}$ m^{-2} s^{-1}			
Irradiation	Decay	Measurement	Analyzed element
5 – 30 seconds	5 – 600 seconds	15 – 300 seconds	Short lived
1 – 8 hours	3 – 5 days	1 – 4 hours	Medium lived
	20 days	1 – 16 hours	Long lived

Table 3. Example of typical ranges of experimental conditions of an INAA procedure.

2.3. Derivation of the measurement equation

The reaction rate R per nucleus capturing a neutron is given by:

$$R = \int_0^\infty \sigma(v)\phi'(v)dv = \int_0^\infty \sigma(E)\phi'(E)dE = \int_0^\infty n'(v)v\sigma(v)dv, \tag{1}$$

where:

σ (v) is the (n,γ) cross section (in cm^2 ; 1 barn (b) = 10^{-24} cm^2) at neutron velocity v (in cm s^{-1});

σ (E) is the (n,γ) cross section (in cm^2) at neutron energy E (in eV);

Φ'(v) is the neutron flux per unit of velocity interval (in cm^{-3}) at neutron velocity v;

n'(v) is the neutron density per unit of velocity interval (in cm^{-4} s) at neutron velocity v;

Φ'(E) is the neutron flux per unit of energy interval (in cm^{-2} s^{-1} eV^{-1}) at neutron energy E.

In Eq.(1), σ (v) = σ (E) with E (in erg = $6.2415.10^{11}$ eV) = ½ m_n v^2 [m_n rest mass of the neutron = 1.6749 10^{-24} g]. Furthermore, per definition, φ'(v) dv = φ'(E)dE (both in cm^{-2} s^{-1}).

In Eq.1, the functions σ(v) [= σ (E)] and φ'(v) [φ'(E)] are complex and are respectively depending on the (n,γ) reaction and on the irradiation site.

In 1987, F De Corte describes in his Aggregate thesis "Chapter 1: fundamentals [3] that the introduction of some generally valid characteristics yields the possibility of avoiding the actual integration and describing accurately the reaction rate in a relatively simple way by means of so-called formalisms or conventions. In short, these characteristics are:

In nuclear research reactors – which are intense sources of neutrons – three types of neutrons can be distinguished. The neutron flux distribution can be divided into three components (see Figure 9):

1. Fission or fast neutrons released in the fission of 235U. Their energy distribution ranges from 100 keV to 25 MeV with a maximum fraction at 2 MeV. These neutrons are slowed down by interaction with a moderator, e.g. H2O, to enhance the probability of them causing a fission chain reaction in the 235U.

2. The epithermal neutron component consists of neutrons (energies from 0.5 eV to about 100 keV). A cadmium foil 1 mm thick absorbs all thermal neutrons but will allow epithermal and fast neutrons above 0.5 eV in energy to pass through. Both thermal and epithermal neutrons induce (n,γ) reactions on target nuclei.

3. The thermal neutron component consists of low-energy neutrons (energies below 0.5 eV) in thermal equilibrium with atoms in the reactor's moderator. At room temperature, the energy spectrum of thermal neutrons is best described by a Maxwell-Boltzmann distribution with a mean energy of 0.025 eV and a most probable velocity of 2200 m/s. In general, a 1 MW reactor has a peak thermal neutron flux of approximately 10^{13} n/cm^2.

The (n,γ) cross section function, σ(v) versus v can be interpreted as a σ(v) ~ 1/v dependence, or σ (E) ~ 1/E$^{1/2}$ dependence [log σ (E) versus log E is linear with slope -1/2], on which (above some eV) several resonances are superposed see Figure 10 taken from http://thor-ea.wikia.com/wiki/Thermal,_Epithermal_and_Fast_Neutron_Spectra web page.

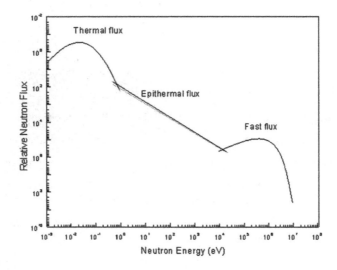

Figure 9. A typical reactor neutron energy spectrum showing the various components used to describe the neutron energy regions.

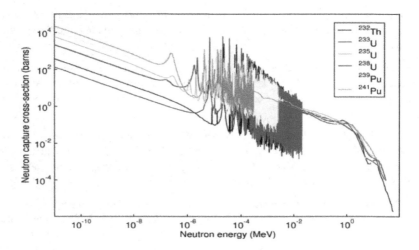

Figure 10. Relation between neutron cross section and neutron energy for major actinides (n, capture).

An NAA technique that employs only epithermal neutrons to induce (n,γ) reactions by irradiating the samples being analyzed inside either cadmium or boron a shield is called epithermal neutron activation analysis (ENAA).

The production of radioactive nuclei is described by:

$$\frac{dN}{dt} = RN_0 - \lambda N \tag{2}$$

In which N_0 number of target nuclei, N is the number of radioactive nuclei, λ is the decay constant in s^{-1}. The disintegration rate of the produced radionuclide at the end of the irradiation time ti follows from:

$$D(t_i) = N(t_i)\lambda = N_0 R(1 - e^{-\lambda t_i}) \tag{3}$$

where:

D is the disintegration rate in Bq of the produced radionuclide, assuming that N=0 at t=0 and N0=constant.

The dependence of the activation cross section and neutron fluence rate to the neutron energy can be taken into account in Eq. (1) by dividing the neutron spectrum into a thermal and an epithermal region; the division is made at En=0.55 eV (the so-called cadmium cut-off energy). This approach is commonly known as the Høgdahl convention [4].

The integral in Eq. (1) can then be rewritten as:

$$R = \int_0^{v_{Cd}} n(v)\,v\,\sigma(v)dv + \int_{v_{Cd}}^{\infty} n(v)v\sigma(v)dv \tag{4}$$

The first term can be integrated straightforward:

$$\int_0^{v_{Cd}} n(v)vdv = v_0\sigma_0 \int_0^{v_{Cd}} n(v)dv = nv_0\sigma_0 \tag{5}$$

in which,

$$n = \int_0^{v_{Cd}} n(v)dv \tag{6}$$

is called the thermal neutron density, with $\Phi_{th} = nv_0$,

- Φ_{th} is the conventional thermal neutron fluence rate, $m^{-2}\ s^{-1}$, for energies up to the Cd cut-off energy of 0.55 eV;

- σ_0 is the thermal neutron activation cross section, m^2, at 0.025 eV;

- v_0 is the most probable neutron velocity at 20 °C: 2200 m s^{-1}.

The second term is re-formulated in terms of neutron energy rather than neutron velocity and the infinite dilution resonance integral I_0 – which effectively is also a cross section (m^2) – is introduced:

$$\int_{v_{Cd}}^{\infty} n(v)v\,dv = \varphi_{epi} \int_{E_{Cd}}^{E_{max}} \frac{\sigma(E_n)dE_n}{E_n} = \varphi_{epi}I_0 \qquad (7)$$

with:

$$I_0 = \int_{E_{Cd}}^{E_{max}} \frac{\sigma(E_n)dE_n}{E_n} \qquad (8)$$

Here, Φ_{epi} the conventional epithermal neutron fluence rate per unit energy interval, at 1 eV.

From this definition of I_0 it can be seen that it assumes that the energy dependency of the epithermal neutron fluence rate is proportional to $1/E_n$. This requirement is fulfilled to a good approximation by most of the (n,γ) reactions.

In the practice of nuclear reactor facilities the epithermal neutron fluence rate Φ_{epi} is not precisely following the inverse proportionality to the neutron energy; the small deviation can be accounted for by introducing an epithermal fluence rate distribution parameter α:

$$I_0(\alpha) = (1\ eV)^{\alpha} \int_{E_{Cd}}^{E_{max}} \frac{\sigma(E_n)dE_n}{E_n^{(1+\alpha)}} \qquad (9)$$

The expression for the reaction rate can thus be re-written as:

$$R = \varphi_{th}\sigma_0 + \varphi_{epi}I_0(\alpha) \qquad (10)$$

Expressing the ratio of the thermal neutron fluence rate and the epithermal neutron fluence rate as $f=\Phi_{th}/\Phi_{epi}$ and the ratio of the resonance integral and the thermal activation cross section as $Q_0(\alpha)= I_0(\alpha)/\sigma_0$, an effective cross section can be defined:

$$\sigma_{eff} = \sigma_0(1 + \frac{Q_0(\alpha)}{f}) \qquad (11)$$

It simplifies the Eq. (10) for the reaction rate to:

$$R = \varphi_{th}\sigma_{eff} \qquad (12)$$

This reaction rate applies to infinite thin objects. In objects of defined dimensions, the inside part will experience a lower neutron fluence rate than the outside part because neutrons are removed by absorption.

The nuclear transformations are established by measurement of the number of nuclear decays. The number of activated nuclei $N(t_i, t_d)$ present at the start of the measurement is given by:

$$N(t_i, t_d, t_m) = \frac{RN_0}{\lambda}(1 - e^{-\lambda t_i})e^{-\lambda t_d} \qquad (13)$$

and the number of nuclei ΔN disintegrating during the measurement is given by:

$$\Delta N(t_i, t_d, t_m) = \frac{RN_0}{\lambda}(1 - e^{-\lambda t_i})e^{-\lambda t_d}(1 - e^{-\lambda t_m}) \qquad (14)$$

in which t_d is the decay or waiting time, i.e. the time between the end of the irradiation and the start of the measurement t_m is the duration of the measurement. Additional correction resulting from high counting rates may be necessary depending upon the gamma-ray spectrometer hardware used as illustrated in chapter 2 [1]. Replacing the number of target nuclei N_0 by $(N_{Av}m)/M$ and using the Eq. (12) for the reaction rate, the resulting net counts C in a peak in the spectrum corresponding with a given photon energy is approximated by the activation formula:

$$N_p = \Delta N\gamma\varepsilon = \varphi_{th}\sigma_{eff}\frac{N_{av}\theta m_x}{M_a}(1 - e^{-\lambda t_i})e^{-\lambda t_d}(1 - e^{-\lambda t_m})I\varepsilon \qquad (15)$$

with:

- N_p is the net counts in the γ-ray peak of E_γ;
- N_{Av} is the Avogadro's number in mol^{-1};
- θ is isotopic abundance of the target isotope;

- m_x is the mass of the irradiated element in g;

- M_a is the atomic mass in g mol^{-1};

- I is the gamma-ray abundance, i.e. the probability of the disintegrating nucleus emitting a photon of E_γ (photons disintegration^{-1});

- ε is the full energy photopeak efficiency of the detector, i.e. the probability that an emitted photon of given energy will be detected and contribute to the photopeak at energy E_γ in the spectrum.

Although the photons emitted have energies ranging from tens of keVs to MeVs and have high penetrating powers, they still can be absorbed or scattered in the sample itself depending on the sample size, composition and photon energy. This effect is called gamma-ray self-attenuation. Also, two or more photons may be detected simultaneously within the time resolution of the detector; this effect is called summation.

Eq. (15) can be simply rewritten towards the measurement equation of NAA, which shows how the mass of an element measured can be derived from the net peak area C:

$$m_x = N_p \cdot \frac{M_a}{N_{av} \cdot \theta} \cdot \frac{\lambda}{\varphi_{th} \cdot \sigma_{eff} \cdot I \cdot \varepsilon \cdot (1 - e^{-\lambda t_i}) \cdot e^{-\lambda t_d} \cdot (1 - e^{-\lambda t_m})} \tag{16}$$

2.4. Standardization

Standardization is based on the determination of the proportionality factors F that relate the net peak areas in the gamma-ray spectrum to the amounts of the elements present in the sample under given experimental conditions:

$$F = \frac{N_p}{M} \tag{17}$$

Both absolute and relative methods of calibration exist.

2.4.1. Absolute calibration

The values of the physical parameters determining the proportionality factor θ, N_{Av}, M, σ_{eff} I, λ, are taken from literature. The parameters σ_{eff} respectively I, λ are not precisely known for many (n,γ) reactions and radionuclides, and in some cases θ is also not accurately known. Since the various parameters were often achieved via independent methods, their individual uncertainties will add up in the combined uncertainty of measurement of the elemental amounts, leading to a relatively large combined standard uncertainty. Moreover, the metrological traceability of the values of the physical constants is not known for all radionuclides. The other parameters N_p, m_x, Φ, ε, t_i, t_d, t_m are determined, calculated or measured for the given circumstances and uncertainties can be established.

2.4.2. Relative calibration

a. Direct comparator method

The unknown sample is irradiated together with a calibrator containing a known amount of the element(s) of interest. The calibrator is measured under the same conditions as the sample (sample-to-detector distance, equivalent sample size and if possible equivalent in composition). From comparison of the net peak areas in the two measured spectra the mass of the element of interest can be calculated:

$$m_{x(unk)} = m_{x(cal)} \cdot \frac{\left(\dfrac{N_p}{t_m \cdot e^{-\lambda t_d} \cdot (1 - e^{-\lambda t_m})} \right)_{unk}}{\left(\dfrac{N_p}{t_m \cdot e^{-\lambda t_d} \cdot (1 - e^{-\lambda t_m})} \right)_{cal}} \tag{18}$$

in which $m_x(unk)$, $m_x(cal)$ mass of the element of interest, in the unknown sample and the calibrator, respectively in g.

In this procedure many of the experimental parameters - such as neutron fluence rate, cross section and photopeak efficiency cancel out at the calculation of the mass and the remaining parameters are all known. This calibration procedure is used if the highest degree of accuracy is required.

The relative calibration on basis of element calibrators is not immediately suitable for laboratories aiming at the full multi-element powers of INAA. It takes considerable effort to prepare multi-element calibrators for all 70 elements measurable via NAA with adequate degree of accuracy in a volume closely matching the size and the shape of the samples. Single comparator method Multi-element INAA on basis of the relative calibration method is feasible when performed according to the principles of the single comparator method. Assuming stability in time of all relevant experimental conditions, calibrators for all elements are co-irradiated each in turn with the chosen single comparator element. Once the sensitivity for all elements relative to the comparator element has been determined (expressed as the so-called k-factor, see below), only the comparator element has to be used in routine measurements instead of individual calibrators for each element. The single comparator method for multi-element INAA was based on the ratio of proportionality factors of the element of interest and of the comparator element after correction for saturation, decay, counting and sample weights defined the k-factor for each element i as:

$$k_i = \frac{(M_a)_{i,cal} \gamma_{comp} \varepsilon_{comp} \theta_{comp} \sigma_{eff,comp}}{M_{i,cal} \gamma_{i,cal} \varepsilon_{i,cal} \theta_{i,cal} (\sigma_{eff})_{i,cal}} \tag{19}$$

Masses for each element i then can be calculated from these k_i factors; for an element determined via a directly produced radionuclide the mass $m_x(unk)$ follows from:

$$m_{x(unk)} = m_{x(comp)} \cdot \frac{\left(\dfrac{N_p}{(1-e^{-\lambda t_m}).t_m.e^{-\lambda t_d}.(1-e^{-\lambda t_m})} \right)_{unk}}{\left(\dfrac{N_p}{(1-e^{-\lambda t_m}).t_m.e^{-\lambda t_d}.(1-e^{-\lambda t_m})} \right)_{comp}} \cdot k_i \qquad (20)$$

where: $m_x(comp)$ is the mass of element x in comparator in g.

These experimentally determined k-factors are often more accurate than when calculated on basis of literature data as in the absolute calibration method. However, the k-factors are only valid for a specific detector, a specific counting geometry and irradiation facility, and remain valid only as long as the neutron fluence rate parameters of the irradiation facility remain stable. The single comparator method requires laborious calibrations in advance, and finally yield relatively (compared to the direct comparator method) higher uncertainties of the measured values. Moreover, it requires experimental determination of the photopeak efficiencies of the detector. Metrological traceability of the measured values to the S.I. may be demonstrated.

b. The k_0-comparator method

The k_0-based neutron activation analysis (k_0-NAA) technique, developed in 1970s, is being increasingly used for multielement analysis in a variety of matrices using reactor neutrons [4-10]. In our research reactor, the k_0-method was successfully developed using the Høgdahl formalism [11]. In the $k0$-based neutron activation analysis the evaluation of the analytical result is based on the so-called k_0- factors that are associated with each gamma-line in the gamma-spectrum of the activated sample. These factors replace nuclear constants, such as cross sections and gamma-emission probabilities, and are determined in specialized NAA laboratories. This technique has been reported to be flexible with respect to changes in irradiation and measuring conditions, to be simpler than the relative comparator technique in terms of experiments but involves more complex formulae and calculations, and to eliminate the need for using multielement standards. The k_0-NAA technique, in general, uses input parameters such as (1) the epithermal neutron flux shape factor (α), (2) subcadmium-to-epithermal neutron flux ratio (f), (3) modified spectral index $r(\alpha)\sqrt{T_n/T_0}$, (4) Westcott's $g(T_n)$-factor, (5) the full energy peak detection efficiency (ε_p), and (6) nuclear data on Q_0 (ratio of resonance integral (I_0) to thermal neutron cross section (σ_0) and k_0. The parameters from (1) to (4) are dependent on each irradiation facility and the parameter (5) is dependent on each counting facility. The neutron field in a nuclear reactor contains an epithermal component that contributes to the sample neutron activation [12]. Furthermore, for nuclides with the Westcott's $g(T_n)$-factor different from unity, the Høgdahl convention should not be applied and the neutron temperature should be introduced for application of a more sophisticated formalism [14], the Westcott formalism. These two formalisms should be taken into account in order to preserve the accuracy of k_0-method.

The k_0-NAA method is at present capable of tackling a large variety of analytical problems when it comes to the multi-element determination in many practical samples. In this part, we have published a paper [15] for which the determination of the Westcott and Høgdahl parameters have been carried out to assess the applicability of the k_0-NAA method using the experimental system and irradiation channels at Es-Salam research reactor.

During the three last decades Frans de Corte and his co-workers focused their investigations to develop a method based on co-irradiation of a sample and a neutron flux monitor, such as gold and the use of a composite nuclear constant called k_0-factor [3, 16]. In addition, this method allows to analyze the sample without use the reference standard like INAA method. The k-factors have been defined as independent of neutron fluence rate parameters as well as of spectrometer characteristics. In this approach, the irradiation parameter $(1+Q_0(\alpha)/f)$ (Eq. (11)) and the detection efficiency ε are separated in the expression (19) of the k-factor, which resulted at the definition of the k_0-factor.

$$k_0 = \frac{1}{k} \cdot \frac{1+Q_{0,comp}(\alpha)/f}{1+Q_{0,cal}(\alpha)/f} \cdot \frac{\varepsilon_{comp}}{\varepsilon_{comp}} = \frac{M_{comp}}{\theta_{comp}\sigma_{0,comp}\gamma_{comp}} \cdot \frac{\theta_{cal}\sigma_{0,cal}\gamma_{cal}}{M_{cal}} \tag{21}$$

$$m_{x(unk)} = m_{x(comp)} \frac{1+Q_{0,comp}(\alpha)/f}{1+Q_{0,cal}(\alpha)/f} \cdot \frac{\varepsilon_{comp}}{\varepsilon_{comp}} \cdot \frac{\left(\dfrac{N_p/t_m}{(1-e^{-\lambda t_i}).e^{-\lambda t_d}(1-e^{-\lambda t_m}).m}\right)_{unk}}{\left(\dfrac{N_p/t_m}{(1-e^{-\lambda t_i}).e^{-\lambda t_d}(1-e^{-\lambda t_m}).m}\right)_{comp}} \cdot \frac{1}{k_0} \tag{22}$$

The applicability of HØGDAHL convention is restricted to (n,γ) reactions for which WEST-COTT's g-factor is equal to unity (independent of neutron temperature), the cases for which WESTCOTT's g = 1 [3, 4, 17], such as the reactions [151]Eu(n, γ) and [176]Lu(n, γ) are excluded from being dealt with. Compared with relative method k_0-NAA is experimentally simpler (it eliminates the need for multi-element standards [3, 18], but requires more complicated calculations [19]. In our research reactor, the k_0-method was successfully developed using the HØGDAHL convention and WESTCOTT formalism [11, 15]. The k_0-method requires tedious characterizations of the irradiation and measurement conditions and results, like the single comparator method, in relatively high uncertainties of the measured values of the masses. Moreover, metrological traceability of the currently existing k_0 values and associated parameters to the S.I. is not yet transparent and most probably not possible. Summarizing, relative calibration by the direct comparator method renders the lowest uncertainties of the measured values whereas metrological traceability of these values to the S.I. can easily be demonstrated. As such, this approach is often preferred from a metrological viewpoint. The concentration of an element can be determined as:

$$\rho_x(ppm) = \frac{\left[\dfrac{N_p/t_m}{SDCW}\right]_x}{\left[\dfrac{N_p/t_m}{SDCW}\right]_{Au}} \cdot \frac{1}{k_{0,Au(x)}} \cdot \frac{G_{th,Au}f + G_{epi,Au}Q_{0,Au}(\alpha)}{G_{th,x}f + G_{epi,x}Q_{0,x}(\alpha)} \cdot \frac{\varepsilon_{p,Au}}{\varepsilon_{p,x}} \times 10^6 \tag{23}$$

Where: the indices x and Au refer to the sample and the monitor, respectively; W_{Au} and W_x represent the mass of the gold monitor and the sample (in g); N_p is the measured peak area, corrected for dead time and true coincidence; S, D, C are the saturation, decay and counting factors, respectively; tm is the measuring time; G_{th} and G_e are the correction factors for thermal and epithermal neutron self shielding, respectively.

2.5. Sources errors

Many publications reported in literature [20-25] treat the concept of evaluation of uncertainties in large range of analytical techniques.

We can give in this part of chapter, the evaluation of uncertainties for neutron activation analysis measurements. Among the techniques of standardization the comparator method for which the individual uncertainty components associated with measurements made with neutron activation analysis (NAA) using the comparator method of standardization (calibration), as well as methods to evaluate each one of these uncertainty components [1].

This description assumes basic knowledge of the NAA method, and that experimental parameters including sample and standard masses, as well as activation, decay, and counting times have been optimized for each measurement. It also assumes that the neutron irradiation facilities and gamma-ray spectrometry systems have been characterized and optimized appropriately, and that the choice of irradiation facility and detection system is appropriate for the measurement performed. Careful and thoughtful experimental design is often the best means of reducing uncertainties. The comparator method involves irradiating and counting a known amount of each element under investigation using the same or very similar conditions as used for the unknown samples. Summarizing, relative calibration by the direct comparator method renders the lowest uncertainties of the measured values whereas metrological traceability of these values to the S.I. can easily be demonstrated. As such, this approach is often preferred from a metrological viewpoint. The measurement equation can be further simplified, by substituting:

$$A_0 = \frac{\lambda N_p e^{\lambda t_d}}{(1 - e^{-\lambda t_m}) \cdot (1 - e^{-\lambda t_i})} \cdot f_p \cdot f_{ltc} \tag{24}$$

in :

$$m_{unk} = m_{cal} \frac{A_{0(unk)}}{A_{0(cal)}} R_\theta R_\Phi R_\sigma R_\varepsilon \text{-blank} \tag{25}$$

Where: R_θ is the ratio of isotopic abundances for unknown sample and calibrator, R_ϕ is the ratio of neutron fluence rates (including fluence gradient, neutron self shielding, and scattering) for unknown sample and calibrator, R_σ is the ratio of effective cross sections if neutron spectrum shape differs from unknown sample to calibrator, R_ε is the ratio of counting efficiencies (differences due to geometry and γ-ray self shielding) for unknown sample and calibrator, blank is the mass of element x in the blank, f_P is the correction for pulse pileup (correction method depends upon the actual hardware used) and f_{ltc} is the correction for inadequacy of live time extension (correction method depends upon the actual hardware used)

Note that the R values are normally very close to unity, and all units are either SI-based or dimensionless ratios. Thus an uncertainty budget can be developed using only SI units and dimensionless ratios for an NAA measurement by evaluating the uncertainties for each of the terms in Eqs. (23) and (24), and for any additional corrections required (e.g., interferences, dry mass conversion factors, etc.).

Uncertainties for some of the terms in Eq. (24) have multiple components. If we sub-divide the uncertainty for each term in the above equations into individual components, add terms for potential corrections, and separate into the four stages of the measurement process, including: pre-irradiation (sample preparation); irradiation; post-irradiation (gamma-ray spectrometry), and radiochemistry, we arrive at the complete list of individual uncertainty components for NAA listed below in Table 4. Only uncertainties from the first three stages should be considered for instrumental neutron activation analysis (INAA) measurements, while all four stages should be considered for radiochemical neutron activation analysis (RNAA) measurements. More details are given in chapter 2 of reference [1] for each subsection of uncertainty component.

1. Pre-irradiation (sample and standard preparation) stage
1.1. Elemental content of standards (comparators)
1.2. Target isotope abundance ratio — unknown samples/standards
1.3. Basis mass (or other sample basis) — including drying
1.4. Sample and standard blanks
2. Irradiation stage
2.1. Neutron fluence exposure differences (ratios) for unknown samples compared to standards (comparators)
2.1. Physical effects (fluence gradients within a single irradiation)
2.2. Temporal effects (fluence variations with time)
2.3. Neutron self shielding (absorption and scattering) effects within a single sample or standard
2.4. Neutron shielding effects from neighbouring samples or standards

2.2. Irradiation interferences

    2.2.1. Fast (high energy) neutron interferences

    2.2.2. Fission interferences

    2.2.3. Multiple neutron capture interferences

2.3. Effective cross section differences between samples and standards

2.4. Irradiation losses and gains

    2.4.1. Hot atom transfer (losses and gains by recoil, nanometer movement)

    2.4.2. Transfer of material through irradiation container

    2.4.3. Sample loss during transfer from irradiation container

    2.4.4. Target isotope burn up differences

2.5. Irradiation timing and decay corrections during irradiation (effects of half life and timing uncertainties)

3. Gamma-ray spectrometry stage

3.1. Measurement replication or counting statistics (depending on number of replicates) for unknown samples

3.2. Measurement replication or counting statistics (depending on number of replicates) for comparator standards

3.3. Corrections for radioactive decay (effects of half life and timing uncertainties for each measurement)

    3.3.1. From end of irradiation to start of measurement

    3.3.2. Effects of clock time uncertainty

    3.3.3. Effects of live time uncertainty

    3.3.4. Count-rate effects for each measurement

        3.3.4.1. Corrections for losses due to pulse pileup for conventional analyzer systems

        3.3.4.2. Effects due to inadequacy of live-time extension for conventional analyzer systems

        3.3.4.3. Uncertainties due to hardware corrections for Loss-Free or Zero Dead Time systems

3.4. Corrections for gamma-ray interferences

3.5. Corrections for counting efficiency differences (if necessary), or uncertainty for potential differences

    3.5.1. Effects resulting from physical differences in size and shape of samples versus standards

    3.5.2. Corrections for gamma-ray self absorption

3.6. Potential bias due to peak integration method

3.7. Potential bias due to perturbed angular correlations (-ray directional effects)

4. Radiochemical stage (only if radiochemical separations are employed)

4.1. Losses during chemical separation

4.2. Losses before equilibration with carrier or tracer

Table 4. Complete list of individual uncertainty components for NAA measurements using the comparator method of standardization; line numbers in this table represent subsections.

2.6. Detection limits of NAA

The detection limit represents the ability of a given NAA procedure to determine the minimum amounts of an element reliably. The detection limit depends on the irradiation, the decay and the counting conditions. It also depends on the interference situation including such things as the ambient background, the Compton continuum from higher energy-rays, as well as any-ray spectrum interferences from such factors as the blank from pre-irradiation treatment and from packing materials. The detection limit is often calculated using Currie's formula:

$$DL = 2.71 + 4.65B , \tag{26}$$

where: DL is the detection limit and B is the background under a gamma-ray peak. This relation is valid only when the gamma-ray background (counting statistical error) is the major interference.

However, practically, the INAA detection limits depend on:

1. The amount of material to be irradiated and to be counted. This is often set by availability, sample encapsulation aspects and safety limits both related to irradiation (irradiation containers) and counting (e.g. with Ge well-type detectors), and possibly because of neutron self-shielding and gamma-ray self-absorption effects. For these reasons practically the sample mass is often limited to approximately 250 mg.

2. The neutron fluxes. These are clearly set by the available irradiation facilities.

3. The duration of the irradiation time. This is set by practical aspects, such as the limitations in total irradiation dose of the plastic containers because of radiation damage. The maximum irradiation time for polyethylene capsules is usually limited to several hours, for instance 5 hours at 5×10^{17} m^{-2}s^{-1}.

4. The total induced radioactivity that can be measured is set by the state-of-the-art of counting and signal processing equipment, with additional radiation dose and shielding considerations. As an example, the maximum activity at the moment of counting may have to be limited to approximately 250 kBq.

5. The duration of the counting time. A very long counting time may set limits to the number of samples processed simultaneously in case the radioactivity decays considerably during this counting time. Moreover, it reduces sample throughput.

6. The total turn-around time. Although sometimes better detection limits may be obtained at long decay times, the demands regarding the turn-around time often imply that a compromise has to be found between the longest permissible decay time and customer satisfaction.

7. The detector size, counting geometry and background shielding. The detector's characteristics may be set in advance by availability but several options exist.

It all illustrates that the detection limit for a given element by INAA may be different for each individual type of material, and analysis conditions. In Table 5 are given, as an indication, typical detection limits as derived from the analysis of a plant and a soil material. Peter Bode in his PhD thesis, Instrumental and organizational aspects of a neutron activation analysis laboratory, the typical detection limits as derived from the analysis of a plant and a soil material given in table 5 [26].

Element	Plant	Soil	Element	Plant	Soil	Element	Plant	Soil
NA	2	10	Nd	0.7	8	Ag	0.2	2
Ca	700	4000	Eu	0.006	0.05	Sn	10	20
Cr	1	1	Yb	0.03	0.2	Te	0.3	3
Co	0.02	0.3	Hf	0.01	0.1	Ba	10	40
Zn	0.4	6	W	0.3	1	Ce	0.2	1
As	0.2	0.8	Os	0.1	0.6	Sm	0.01	0.03
Br	0.3	0.8	Au	0.003	0.01	Tb	0.008	0.1
Sr	5	60	Th	0.01	0.1	Lu	0.004	0.02
Mo	4	10	K	200	1500	Ta	0.01	0.2
Cd	3	8	Sc	0.001	0.02	Re	0.08	0.2
Sb	0.02	0.2	Fe	8	100	Ir	0.0006	0.004
Cs	0.02	0.3	Ni	2	30	Hg	0.05	0.4
La	0.1	0.3	Ga	2	10	U	0.2	2
Se	0.1	1	Rb	0.4	6	Zr	5	80

Table 5. Detection limits of elements in $mg.kg^{-1}$ as observed in NAA procedure of plant material and a soil material.

3. Applications

It is hardly possible to provide a complete survey of current NAA applications; however, some trends can be identified [27]. At specialized institutions, NAA is widely used for analysis of samples within environmental specimen banking programmes [28]. The extensive use of NAA in environmental control and monitoring can be demonstrated by the large number of papers presented at two symposia organized by the IAEA in these fields: "Applications of Isotopes and Radiation in Conservation of the Environment" in 1992 [29] and "Harmonization of Health-Related Environmental Measurements Using Nuclear and Isotopic Techniques" in 1996 [30]. Similar trends can also be identified from the topics discussed at the regular conference on "Modern Trends in Activation Analysis (MTAA)" and at the symposia on "Nuclear Analytical Methods in the Life Sciences" [31-33]. Additional sources of recent information on utilizing NAA in selected fields, such as air pollution and environmental analysis, food, forensic science, geological and inorganic materials as well as water analysis can be found in the bi-annual reviews in Analytical Chemistry, for instance cf. Refs [34-42]. It follows from these reviews that NAA has been applied for determining many elements, usually trace elements, in the following fields and sample types:

1. Archaeology: samples and objects such as amber, bone, ceramics, coins, glasses, jewellery, metal artefacts and sculptures, mortars, paintings, pigments, pottery, raw materials, soils and clays, stone artefacts and sculptures can be easily analyzed by NAA.

2. Biomedicine: the samples and objects that can be analysed include: animal and human tissues activable tracers, bile, blood and blood components, bone, brain cell components and other tissues, breast tissue, cancerous tissues, colon, dialysis fluids, drugs and medicines, eye, faeces, foetus, gallstones, hair, implant corrosion, kidney and kidney stones, liver, lung, medical plants and herbs, milk, mineral availability, muscle, nails, placenta, snake venom, rat tissues (normal and diseased), teeth, dental enamel and dental fillings, thyroid, urine and urinary stones.

In this work, we have used the INAA technique to analyse the traditional medicinal seeds prescribed for specific treatment purposes, were purchased from local markets [43]. The samples were irradiated at Es-Salam research reactor, at a power of 5 MW for 6 h. The accuracy of the method was established by analyzing reference materials. Twenty elements were measured, with good accuracy and reproducibility (Table 6[1]). The concentration of elements determined, was found to vary depending on the seeds (Fig.11). The daily intake of essential and toxic elements was determined, and compared with the recommended values. The probable cumulative intake of toxic elements is well below the tolerance limits.

Element	Unity	Black seeds	Fenugeek	Caraway
Ba	mg/Kg	7.7 ± 5.5	100.3 ± 5.8	112.4 ± 6.5
Br	mg/Kg	136.9 ± 4.6	119.6 ± 3.9	72.9 ± 2.4
Ca	g/Kg	3.77 ± 4.55	3.14 ± 0.39	1.50 ± 0.21
Ce	mg/Kg	1.44± 0.07	2.6 ± 0.1	1.98 ± 0.11
Co	mg/Kg	0.66 ± 0.02	0.73 ± 0.02	0.81 ± 0.03
Cr	mg/Kg	4.44 ± 0.19	29.3 ± 1.0	2.96 ± 0.20
Cs	mg/Kg	0.25 ± 0.01	0.51 ± 0.02	0.22 ± 0.01
Eu	mg/Kg	0.022 ± 0.002	0.039 ± 0.002	0.023 ± 0.002
Fe	mg/Kg	656.2 ± 71.6	823.2 ± 89.8	674.67 ± 74.16
K	g/Kg	3.67± 1.79	3.75 ± 0.20	3.7 ± 0.2
La	mg/Kg	0.74 ± 0.04	1.53 ± 0.06	1.50 ± 0.06
Na	mg/Kg	1028 ± 34	804.20 ± 26.69	615.50 ± 20.41
Rb	mg/Kg	24 ± 2	36.8 ± 1.4	26.3 ± 1.9
Sc	mg/Kg	0,258 ± 0,037	0,362 ± 0,051	0,272 ± 0,008
Se	mg/Kg	0,29 ± 0,04	ND	ND

---

1 nd : not detected.

Element	Unity	Black seeds	Fenugeek	Caraway
Sm	mg/Kg	0,092 ± 0,004	0,18 ± 0,01	0,142 ± 0,005
Sr	mg/Kg	203,2 ± 7,8	136,88 ± 7,4	101,7 ± 4,7
Th	mg/Kg	0,159 ± 0,009	0,32 ± 0,02	0,195 ± 0,014
Zn	mg/Kg	68,06 ± 2,11	42,8 ± 1,4	40,24 ± 1,30

Table 6. Elemental concentrations in the medicinal seed samples (Black seeds, Fenugreek, Caraway).

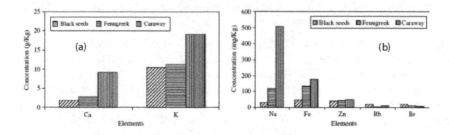

Figure 11. Concentration of the major and minor (a) and trace (b) elements in the medicinal seed samples.

3. Environmental: in this domain, related fields concerned by NAA are: aerosols, atmospheric particulates (size fractionated), dust, fossil fuels and their ashes, flue gas, animals, birds, insects, fish, aquatic and marine biota, seaweed, algae, lichens, mosses, plants, trees (leaves, needles, tree bark), household and municipal waste, rain and horizontal precipitations (fog, icing, hoarfrost), soils, sediments and their leachates, sewage sludges, tobacco and tobacco smoke, surface and ground waters, volcanic gases.

Recently, our laboratory is strongly involved in various areas of application of k_0-NAA. The present work focuses on the application of the k_0-NAA method in Nutritional and Health-Related Environmental field [44]. Tobacco holds a leading position among different commodities of human consumption. The adverse health effects of toxic and trace elements in tobacco smoke on smokers and non-smokers are a special concern. In the present study, the concentration of 24 trace elements in cigarette tobacco of five different brands of Algerian and American cigarettes have been determined by k_0-based INAA method. The results were compared with those obtained for samples from Iranian, Turkish, Brazilian and Mexican cigarettes tobacco. To evaluate the accurate of the results the SRM IAEA-140/TM was executed.

A multi-element analysis procedure based on the k_0-NAA method was developed at Es-Salam research reactor allowing to simultaneously determine concentrations for 24 elements (As, Ba, Br, Ca, Ce, Co, Cr, Cs, Eu, Fe, Hf, K, La, Na, Rb, Sb, Sc, Se, Sm, Sr, Ta,

Tb, Th, Zn). The determination of toxic and trace elements in cigarette tobacco is important both from the point of view of health studies connected with smoking and more general aspects of the uptake of trace elements by plants (table 7). Because of its great sensitivity, k_0-NAA method is very suitable for determination of heavy metals such as As, Sb, Se and Zn. The accuracy of the results was checked by the analysis of standard reference material and good agreement was obtained with certified or literature values. The results of Algerian tobacco (table 8) were compared with analyses of Turkey [45], Iran [46], Mexican [47] and Brazilian tobacco [48].

Element	Designated sample code				
	T1	T2	T3	T4	T5
As	4.05 ± 0.16	6.4 ± 0.24	2.16 ± 0.09	2.42 ± 0,09	4.27 ± 0.28
Ba	101.3 ± 5.5	100.3 ± 5.8	112.4 ± 6.5	83.1 ± 4.4	120.0 ± 7.7
Br	136.9 ± 4.6	119.6 ± 3.9	72.9 ± 2.4	54.1 ± 1.7	68.7 ± 2.3
Ca %	3.77 ± 4.55	3.14 ± 0.39	1.50 ± 0.21	2.11 ± 0.26	2.39 ± 0.31
Ce	1.44 ± 0.07	2.6 ± 0.1	1.98 ± 0.11	1.01 ± 0.05	1.81 ± 0.11
Co	0.66 ± 0.02	0.73 ± 0.02	0.81 ± 0.03	0.51 ± 0.02	0.78 ± 0.03
Cr	4.44 ± 0.19	29.3 ± 1.0	2.96 ± 0.20	2.37 ± 0.11	2.80 ± 0.22
Cs	0.25 ± 0.01	0.51 ± 0.02	0.22 ± 0.01	0.191 ± 0.008	0.42 ± 0.02
Eu	0.022 ± 0.002	0.039 ± 0.002	0.023 ± 0.002	0.021 ± 0.001	0.032 ± 0.002
Fe	656 ± 72	823 ± 90	675 ± 74	384 ± 13	664 ± 24
Hf	0.127 ± 0.007	0.224 ± 0.015	0.24 ± 0.02	0.144 ± 0.014	0.143 ± 0.010
K %	3.67± 1.79	3.75 ± 0.20	3.7 ± 0.2	2.92 ± 0.14	3.38 ± 0.18
La	0.74 ± 0.04	1.53 ± 0.06	1.50 ± 0.06	1.082 ± 0.043	1.68 ± 0.06
Na	1028 ± 34	804 ± 27	616 ± 20	653 ± 22	575 ± 19
Rb	24 ± 2	36.8 ± 1.4	26.3 ± 1.9	14.33 ± 0.59	25.4 ± 1.6
Sb	0.089 ± 0.014	0.180 ± 0.020	0.036 ± 0.007	0.127 ± 0.015	0.346 ± 0.037
Sc	0.258 ± 0.037	0.362 ± 0.051	0.272 ± 0.008	0.165 ± 0.023	0.264 ± 0.037
Se	0.29 ± 0.04	ND	ND	ND	ND
Sm	0.092 ± 0.004	0.18 ± 0.01	0.142 ± 0.005	0.095 ± 0.004	0.152 ± 0.005

Element	Designated sample code				
	T1	T2	T3	T4	T5
Sr	203.2 ± 7.8	136.9 ± 7.4	101.7 ± 4.7	82.37 ± 3.62	106.80 ± 5.39
Ta	0.021 ± 0.003	0.043 ± 0.006	0.023 ± 0.004	0.029 ± 0.004	0.023 ± 0.004
Tb	0.018 ± 0.004	0.021 ± 0.003	0.014 ± 0.004	0.008 ± 0.002	0.018 ± 0.004
Th	0.159 ± 0.009	0.32 ± 0.02	0.195 ± 0.014	0.153 ± 0.012	0.19 ± 0.02
Zn	68.06 ± 2.11	42.8 ± 1.4	40.24 ± 1.30	27.53 ± 0.89	42.99 ± 1.38

Table 7. Concentration values (mg kg-1) of five brands of tobacco by k_0-NAA method.

Element	Algeria	Turkey	Iran	Mexican	Brazilian
As	4.05 – 6.4	1.0	-	< 0.55 – 3.24	-
Ba	100.3 – 101.3	64.6	1.15 ± 0.01	64 – 251	45.8 – 99.7
Br	119.6 – 136.9	59.2	137 ± 2	49 -136	-
Ca %	3.14 – 3.77	-	-	-	-
Ce	1.44 – 2.6	-	1.19 ± 0.02	< 0.3 -1.7	1.2 – 8.3
Co	0.66 – 0.73	0.77	0.21 ± 0.01	0.29 – 0.55	0.70 – 1.2
Cr	4.44 – 29.3	7.1	3.14 ± 0.14	< 0.8 – 2.4	-
Cs	0.25 – 0.51	0.20	0.17 ± 0.02	0.091 – 0.4	-
Eu	0.022 – 0.039	0.034	0.045 ± 0.010	< 0.03	0.036 -0.037
Fe	656.2 – 823.2	1000	649 ± 6	420 – 680	710 – 4100
Hf	0.127 – 0.224	0.15	0.09 ± 0.01	< 0.03 – 0.13	0.15 – 0.87
K %	3.67 – 3.75	2.6	0.220 ± 0.005	1.83 – 4.03	-
La	0.74 – 1.53	1.0	0.62 ± 0.02	< 0.2 – 0.66	1.9 – 4.8
Na	804.2 – 1028	394	347 ± 18	309 – 566	-
Rb	24.0 - 36.8	17.7	22.6 ± 3.6	19 – 50	30.3 – 45.0
Sb	0.089 – 0.180	0.10	0.68 ± 0.03	< 0.7	0.079 – 0.47
Sc	0.258 – 0.362	0.35	0.43 ± 0.03	0.13 – 0.22	0.23 – 1.50

Element	Algeria	Turkey	Iran	Mexican	Brazilian
Se	0.29	0.18	-	< 0.7	-
Sm	0.092 – 0.180	-	0.88 ± 0.01	0.07 – 0.14	0.24 – 0.60
Sr	136.88 – 203.20	-	-	227 – 472	-
Ta	0.021 – 0.043	-	-	-	-
Tb	0.018 – 0.021	-	0.034 ± 0.002	-	-
Th	0.159 – 0.320	0.32	0.177 ± 0.010	< 0.1 – 0.17	0.34 – 4.00
Zn	42.80 – 68.06	35	12.6 ± 0.4	14 - 56	-

Table 8. Comparison between our results (Algerian cigarettes tobacco) and those reported in the literature.

4. Forensics: bomb debris, bullet lead, explosives detection, glass fragments, paint, hair, gunshot residue swabs, shotgun pellets.

5. Geology and geochemistry: asbestos, bore hole samples, bulk coals and coal products, coal and oil shale components, crude oils, kerosene, petroleum, cosmo-chemical samples, cosmic dust, lunar samples, coral, diamonds, exploration and geochemistry, meteorites, ocean nodules, rocks, sediments, soils, glacial till, ores and separated minerals.

6. Industrial products: alloys, catalysts, ceramics and refractory materials, coatings, electronic materials, fertilizers, fissile material detection and other safeguard materials, graphite, high purity and high-tech materials, integrated circuit packing materials, online, flow analysis, oil products and solvents, pharmaceutical products, plastics, process control applications, semiconductors, pure silicon and silicon processing, silicon dioxide, NAA irradiation vials, textile dyes, thin metal layers on various substrates.

7. Nutrition: composite diets, foods, food colours, grains, honey, seeds, spices, vegetables, milk and milk formulae, yeast. In this chapter, we focus on the application of the k_0 method of instrumental neutron activation analysis in Nutritional and Health-Related Environmental field [49]. Three kinds of milk were purchased in the powder form from local supermarket. The samples of milk powder were analyzed using k_0-NAA method. Concentrations of six elements Br, Ca, K, Na, Rb and Zn have been determined by long irradiation time with a thermal and epithermal flux of $4.7.10^{12}$ n.cm^{-2}.s^{-1} and $2.29.10^{11}$ n.cm^{-2}.s^{-1}, respectively (see table 9). The reactor neutron spectrum and detection efficiency calibration parameters such as α, f and εp have been used for the calculation of elemental concentrations. The analytical results for three kinds of milk using k_0-NAA are compared with the certified values of SRMs. In this work, we have determined six elements in three kinds of milk and two reference materials, IAEA-153 and IAEA-155. The elements Br, Ca, K, Na, Rb and Zn were determined in each kind of the three samples of milk.

Element	Element Designated sample code		
	M1	M2	M3
Br	12.74 ± 0.81	38.9 ± 1.26	72.73 ± 0.56
Ca	9040 ± 150 (9300)	9560 ± 820(9600)	9210 ± 130
K	12970 ± 780 (12000)	12700 ± 700 (12600)	12700 ± 500
Na	280 ± 100 (3500)	33900 ± 100 (4400)	41400 ± 130
Rb	12.03 ± 0.08	14.7± 0.7	8.2 ± 0.8
Zn	48.03 ± 0.08	42.7 ± 0.2	42.4 ± 0.2

Table 9. Concentration values of Milk: M1, M2 and M3; units are in mg/kg, NB: (value) is the concentration value of indicated by producer.

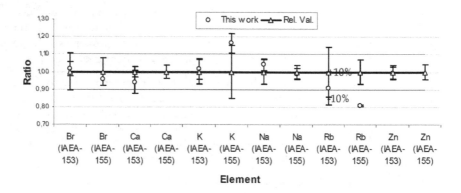

Figure 12. Comparison of k_0-NAA data to certified values for IAEA-153 and IAEA-155.

The accuracy of the measurements was evaluated by analyzing two SRMs Whey powder AIEA-155 and Milk powder AIEA-153. The analysis results illustrated in figure 12 showed that the deviations between experimental and certified values were mostly less than 10%.

As an example, an investigation in the nutrition field was carried out by the radiochemical neutron activation analysis to the proportioning of iodine in food salt [50].

8.   Quality assurance: this include analysis of reference materials, certification of element contents and homogeneity testing of mainly biological and environmental reference materials of chemical composition and methods inter-comparisons. Additional information about these applications can be found in the Proceedings of the Int. Symposia on Biological and Environmental Reference Materials (BERM). In 2012, Hamidatou L et all

reported "k0-NAA quality assessment in an Algerian laboratory by analysis of SMELS and four IAEA reference materials using Es-Salam Research reactor" the internal quality control of the k0-NAA technique [51]. The concept of QC/QA, internal and external validation is considered as an advanced stage in the life cycle of an analytical method.

Our contribution in this domain is considered as periodic activities. Since the Nineties our laboratory participated through AFRA/AIEA projects in different inter-laboratory proficiency tests. Recently, our laboratory was participated in four inter-comparison tests organized by IAEA within the framework of the AFRA project to assess the analytical performance of 18 analytical laboratories participating in the RAF /4/022 project, Enhancement of Research Reactor Utilization and Safety by taking part in analytical proficiency testing IAEA in conjunction with WEPAL, the Wageningen Evaluating Programs for Analytical Laboratories. The Proficiency Testing tests related to the determination of major, minor and trace elements in materials of the International Soil and Plant Analytical Exchange material (Wepal codes ISE, IPE).

9. Neutron flux characterization: theoretical and experimental study, calibration of irradiation channels, simulations using Monte Carlo Code. In general, the implementation of new techniques based on the neutron beams or flux around the research reactors needs the knowledge of the essential parameters of neutron flux in different sites to obtain a better precision during the development.

In this context, we give a great interest in the neutron study for our irradiation channels by making periodic calibrations using experimental and simulation approaches [11, 15, 52].

4. Conclusions

NAA plays a complementary role in materials analysis in an industrial analytical laboratory. There are applications where it is highly desirable, and may play the dominant role as the method of choice e.g. bulk analysis of Si. The advantages of NAA are still the minimum sample preparation and ultra high sensitivity while turnaround time and lack of spatial resolution is a significant limitation.

The many diverse applications in varied fields show that NAA is extremely useful, even though it is a relatively simple analytical method. Irradiation of samples may be done at nuclear reactors that offer such services. Such centres are easily accessible nowadays. Even if these centres are inaccessible, other neutron sources that emit thermal neutrons may be used. Hence any laboratory that has a gamma counter can perform NAA experiments. Development of research programs based on NAA is accessible to any laboratory that is willing to invest a minimal amount of funds. In this chapter, we have presented the neutron activation analysis in different angles such as: basis principles, derivation of several equations, techniques, procedures, etc. In addition, we have associated in each part of general work, some examples of our technical developments and applications of NAA method in several fields. All analytical works were executed in our centre using irradiation facilities of

Es-Salam research reactor and all necessary equipments installed in the NAA department to cover all steps of analytical process.

Acknowledgments

Thanks are due to Dr Derdour Mohamed the responsible of COMENA and Mr Kerris Abdelmoumen General Director of CRNB for financial support. Grateful acknowledgment is made to Mr Salhi Mhamed the director of techniques and nuclear applications division for his highly valuable assistance. Special thanks are due to all colleagues involved for their help during fifteen years.

Author details

Lylia Hamidatou, Hocine Slamene, Tarik Akhal and Boussaad Zouranen

Department of Neutron Activation Analysis, Nuclear Research Centre of Birine, Algeria

References

[1] Greenberg, R.R., et al., Spectrochimica Acta Part B, 66 issues 3-4; 2011, p193-241.

[2] Bé, M. M., Chisté. V., BNM-LNHB/CEA- Tables des radionucléides, 01/12/01-6/4/2004 ; 2004.

[3] De Corte, F.,The k_0 Standardization Method: A Move to the Optimization of NAA, Gent University; 1987.

[4] De Corte, F., et al.,J. Radioanal. Nucl. Chem., volume 197; 1994, p93.

[5] Acharya, R., Chatt, A.,. J. Radioanal. Nucl. Chem., volume 257; 2003, p525.

[6] Simonits, A., De Corte, F., Hoste, J., J. Radioanal. Chem., volume 24; 1975, p. 31.

[7] Lin, X., Li, X., J. Radioanal. Nucl. Chem., volume 223; 1997, p. 47.

[8] Moens, L., De Corte, F., Simonits, A., De Wilspelaere, A., Hoste, J., J. Radioanal. Chem., volume 52 ; 1979, p. 379.

[9] De Corte, F., Simonits, A., De Wilspelaere, A., Hoste, J., J. Radioanal. Nucl. Chem., volume 113; 1987, p145.

[10] Jovanovic, S., De Corte, F., Moens, L., Simonits, A., Hoste, J., J. Radioanal. Nucl. Chem., volume 82; 1984, p.379.

[11] Alghem, L., Ramdhane ,M., S. Khaled, T. Akhal., The development and application of k_0-standardization method of neutron activation analysis at Es-Salam research reactor, Nucl. Instr. and methods., 556; 2006, p386–390.

[12] Mustra, C. O., Freitas, M. C., Alimeida, S. M. (). J. Radioanal. Nucl. Chem. volume 257; 2003, p539.

[13] HyperLab 2005 System: Installation and Quick Start Guides. HyperLabs Software, Budapest, Hungary; 2005.

[14] De Corte, F, Simonits, A, De Wispelaere, A., Hoste, J., Accuracy and applicability of the k_0-standardozation method. J Radioanal Nucl Chem volume 113; 1987,. p145–154.

[15] Alghem Hamidatou, L., Ramdhane, M., Characterization of neutron spectrum at Es-Salam Research Reactor using Høgdahl convention and Westcott formalism for the k0-based neutron activation analysis, Journal of Radioanalytical and Nuclear volume 278 issue (3); 2008, p627-630.

[16] Dung, H.M., Hien, P.D., J. Radioanal. Chem. volume 257; 2003, p643.

[17] De Corte, F., Simonits, A., Bellemanns, F, Freitas, MC., Jovanovic, S, Smodis, B., Erdtmann G., Petri, H., De Wispelaere, A., Recent advances in the k_0-standardization of neutron activation analysis: Extensions, applications, prospects. J Radioanal Nucl Chem volume 169; 1993, p125–158.

[18] Kennedy, G., ST-Pierre, J., Wang, K., Zhang, Y., Preston, J., Grant, C., Vutchkov, M., Activation constants for Slowpoke and MNS Reactors calculated from the neutron spectrum and k_0 and Q_0 values. J Radioanal Nucl Chem volume 245; 2000, p167–172.

[19] Acharya, RN., Nair, AGC., Reddy, AVR., Manohar, SB. () Validation of a neutron activation analysis method using k_0-standardization. Appl. Radiation Isotopes, volume 57; 2002, p391–398.

[20] Kragten, J., Calculating standard deviations and confidence intervals with a universally applicable spreadsheet technique, Analyst volume 119; 1994, p2161–2166.

[21] Rees, C.E., Error propagation calculations, Geochim. Cosmochim. Acta volume 48; 1984

[22] Ellison, S.L.R., Rosslein, M., Williams, A., Quantifying Uncertainty in Analytical Measurements (2nd ed.; QUAM:2000.P1), Eurachem/CITAC; 2000.

[23] Robouch, P., Arana, G., Eguskiza, M., Pommé, S., Etxebarria, N., Uncertainty budget for k0-NAA, J. Radioanal. Nucl. Chem. volume 245; 2000, p195–197.

[24] Joint Committee for Guides in Metrology, Evaluation of measurement data – Supplement 1 to the "Guide to the expression of uncertainty in measurement" – Propagation of distributions using a Monte Carlo method JCGM101:2008 (ISO/IEC Guide 98-3-1); 2008.

[25] Joint Committee for Guides in Metrology, Evaluation of measurement data — Guide to the expression of uncertainty in measurement (2008) JCGM 101:2008 (ISO/IEC Guide 98-3); 2008.

[26] Bode, P., Instrumental and organizational aspects of a neutron activation analysis laboratory Ph.D. dissertation, Delft University of Technology, Delft; 1996.

[27] IAEA TEC-DOC, 1215; 2001.

[28] Rossbach, M., Emons, H., Groemping, A., Ostapczuk, P., Schladit, J.D., Quality control strategies at the environmental specimen bank of the Federal Republic of Germany, Harmonization of Health-Related Environmental Measurements Using Nuclear and Isotopic Techniques, (Proc. Symp. Hyderabad, India, 1996), IAEA, Vienna; 1997, p89–100.

[29] International Atomic Energy Agency, Applications of Isotopes and Radiation in Conservation of the Environment (Proc. Symp. Karlsruhe, Germany, 1992), IAEA, Vienna; 1992.

[30] International Atomic Energy Agency, Harmonization of Health-Related Environmental Measurements Using Nuclear and Isotopic Techniques, (Proc. Symp. Hyderabad, India, 1996), IAEA, Vienna; 1997.

[31] Zeisler, R., Guinn, V.P., Nuclear Analytical Methods in the Life Sciences, Biol. Trace El. Res., 26/27, Humana Press, Clifton, New Jersey; 1990.

[32] Kucera, J., Obrusnik, I., Sabbioni, E., Nuclear Analytical Methods in the Life Sciences, Biol. Trace El. Res., 43/45, Humana Press, Totowa, New Jersey; 1994.

[33] Chai, Z., (Ed.), Nuclear Analytical Methods in the Life Sciences, (1999). Biol. Trace El. Res.71, Humana Press, Totowa, New Jersey.

[34] Ehmann, W.D., Yates, S.W.Nuclear and radiochemical analysis, Anal. Chem. 58 49R-65R, 1986.

[35] Ehmann, W.D., Yates, S.W., Nuclear and radiochemical analysis, Anal. Chem. 60 42R–62R; 1988.

[36] Ehmann, W.D., Robertson, J.D., Yates, S.W., Nuclear and radiochemical analysis, Anal. Chem. 62 5OR–70R; 1990.

[37] Ehmann, W.D., Robertson, J.D., Yates, S.W., Nuclear and radiochemical analysis, Anal. Chem. 64 1R–22R; 1992.

[38] Ehmann, W.D., Robertson, J.D., Yates, S.W., Nuclear and radiochemical analysis, Anal. Chem. 66 229R–251R; 1994.

[39] Application reviews, Analytical Chemistry 63, No. 12 1R-324R; 1991.

[40] Application reviews, Analytical Chemistry 65, No. 12 1R-484R; 1993.

[41] Application reviews, Analytical Chemistry 67, No. 12 1R-583R; 1995.

[42] Application reviews, Analytical Chemistry 69, No. 12 1R-328R; 1997.

[43] Khaled, S., Mouzai, M., Ararem, A., Hamidatou, L., Zouranen, B., Elemental analysis of traditional medicinal seeds by instrumental neutron activation analysis. J Radioanal Nucl Chem volume 281; 2009, p87–90.

[44] Hamidatou, L. A,. Khaled, S., Akhal, T., Ramdhane, M., Determination of trace elements in cigarette tobacco with the k_0-based NAA method using Es-Salam research reactor. J Radioanal Nucl Chem, volume 281; 2009, p535–540

[45] Gülovali, M.C., Gunduz, G., J. Radioanal. Chem., volume 78; 1983, p189-198.

[46] Abedinzadeh, Z., Razechi, M., Parsa, B., J. Radioanal. Chem. volume 35, 1977,p373-379.

[47] Vega-Carrillo, H.R., Iskander, F.Y., Manzanres-Acuna, E., J. Radioanal. Nucl. Chem. Lett., volume 200; 1995, p137-145.

[48] Munita, C.S., Mazzillil, B.P., J. Radioanal. Nucl. Chem. Lett, volume 108; 1986, p217-227.

[49] Hamidatou, L. A., Khaled, S., Mouzai, M., Zouranen, B., Ararem, A., Alghem, A., Ramdhane, M., Instrumental neutron activation analysis of milk samples using the k_0-standardization method at Es-Salam research reactor. Phys. Chem. News volume 45; 2009, p44-47.

[50] Akhal, T., Mouzai, M., Sana, Y., Ladjel, M., Application of RNAA to the proportioning of iodine in food salt. Phys. Chem. News volume 45; 2009, p34-38.

[51] Hamidatou, L. A., S. Dakar, S. Boukari, k0-NAA quality assessment in an Algerian laboratory by analysis of SMELS and four IAEA reference materials using Es-Salam Research reactor, Journal of Nuclear Instruments and methods in physics research section A., 682 (2012) p 75–78.

[52] Hamidatou, L., Benkharfia, H Experimental and MCNP calculations of neutron flux parameters in irradiation channel at Es-Salam reactor, J. Radioanal. Nucl. Chem., V287, N° 3 (2011) p971-975.

Permissions

The contributors of this book come from diverse backgrounds, making this book a truly international effort. This book will bring forth new frontiers with its revolutionizing research information and detailed analysis of the nascent developments around the world.

We would like to thank Dr. Faycal Kharfi, for lending his expertise to make the book truly unique. He has played a crucial role in the development of this book. Without his invaluable contribution this book wouldn't have been possible. He has made vital efforts to compile up to date information on the varied aspects of this subject to make this book a valuable addition to the collection of many professionals and students.

This book was conceptualized with the vision of imparting up-to-date information and advanced data in this field. To ensure the same, a matchless editorial board was set up. Every individual on the board went through rigorous rounds of assessment to prove their worth. After which they invested a large part of their time researching and compiling the most relevant data for our readers. Conferences and sessions were held from time to time between the editorial board and the contributing authors to present the data in the most comprehensible form. The editorial team has worked tirelessly to provide valuable and valid information to help people across the globe.

Every chapter published in this book has been scrutinized by our experts. Their significance has been extensively debated. The topics covered herein carry significant findings which will fuel the growth of the discipline. They may even be implemented as practical applications or may be referred to as a beginning point for another development. Chapters in this book were first published by InTech; hereby published with permission under the Creative Commons Attribution License or equivalent.

The editorial board has been involved in producing this book since its inception. They have spent rigorous hours researching and exploring the diverse topics which have resulted in the successful publishing of this book. They have passed on their knowledge of decades through this book. To expedite this challenging task, the publisher supported the team at every step. A small team of assistant editors was also appointed to further simplify the editing procedure and attain best results for the readers.

Our editorial team has been hand-picked from every corner of the world. Their multi-ethnicity adds dynamic inputs to the discussions which result in innovative outcomes. These outcomes are then further discussed with the researchers and contributors who give their valuable feedback and opinion regarding the same. The feedback is then collaborated with the researches and they are edited in a comprehensive manner to aid the understanding of the subject.

Apart from the editorial board, the designing team has also invested a significant amount of their time in understanding the subject and creating the most relevant covers. They scrutinized every image to scout for the most suitable representation of the subject and create an appropriate cover for the book.

The publishing team has been involved in this book since its early stages. They were actively engaged in every process, be it collecting the data, connecting with the contributors or procuring relevant information. The team has been an ardent support to the editorial, designing and production team. Their endless efforts to recruit the best for this project, has resulted in the accomplishment of this book. They are a veteran in the field of academics and their pool of knowledge is as vast as their experience in printing. Their expertise and guidance has proved useful at every step. Their uncompromising quality standards have made this book an exceptional effort. Their encouragement from time to time has been an inspiration for everyone.

The publisher and the editorial board hope that this book will prove to be a valuable piece of knowledge for researchers, students, practitioners and scholars across the globe.

List of Contributors

Faycal Kharfi
Department of Physics, Faculty of Science, University of Ferhat Abbas-Sétif, Algeria

Samuel Opoku, William Antwi and Stephanie Ruby Sarblah
Department of Radiography, College of Health Sciences, SAHS, University of Ghana, Legon, Accra, Ghana

Mariluce Gonçalves Fonseca
Federal University of Piaui, School of Medicine, UNESP, Botucatu, Brazil

Faycal Kharfi
Department of Physics, Faculty of Science, University of Ferhat Abbas-Sétif, Algeria

Selma Kadioglu
Ankara University, Faculty of Engineering, Department of Geophysical Engineering, Ankara, Turkey
Ankara University, Earth Sciences Application and Research Center, Ankara, Turkey

A.C. Avelar, W.M. Ferreira and M.A.B.C. Menezes
DZOO Department of Animal Sciences, Universidade Federal de Minas Gerais Avenida Antonio Carlos, Campus UFMG, Pampulha, Belo Horizonte, Brazil

Lylia Hamidatou, Hocine Slamene, Tarik Akhal and Boussaad Zouranen
Department of Neutron Activation Analysis, Nuclear Research Centre of Birine, Algeria

Printed in the USA
CPSIA information can be obtained
at www.ICGtesting.com
JSHW011400221024
72173JS00003B/357